Chemical Engineering and the Law Forum 2014

Core Programming Area at the 2014 AIChE Spring Meeting and 10th Global Congress on Process Safety

New Orleans, Louisiana, USA
30 March - 3 April 2014

ISBN: 978-1-63439-067-5

Chemical Engineering and the Law Forum 2014

Core Programming Area at the 2014 AIChE Spring Meeting and 10th Global Congress on Process Safety

New Orleans, Louisiana, USA
30 March - 3 April 2014

ISBN: 978-1-63439-067-5

AIChE

Printed from e-media with permission by:

Curran Associates, Inc.
57 Morehouse Lane
Red Hook, NY 12571

Some format issues inherent in the e-media version may also appear in this print version.

Printed by Curran Associates, Inc. (2014)

For permission requests, please contact AIChE
at the address below.

AIChE
3 Park Avenue
New York, NY 10016-5991

Phone: (203) 702-7660
Fax: (203) 775-5177

www.aiche.org

Additional copies of this publication are available from:

Curran Associates, Inc.
57 Morehouse Lane
Red Hook, NY 12571 USA
Phone: 845-758-0400
Fax: 845-758-2634
Email: curran@proceedings.com
Web: www.proceedings.com

TABLE OF CONTENTS

354012 Possible vs. Practical: Engineers Must Lead the Development of Practical Technologies

Abstract:

Chemical Engineering is need today more than ever. There is a strong societal desire for sources of energy and feedstocks that can sustainably support a high standard of living for all. Thermodynamic, material and economic challenges stand in the way. Most of the general public does not have the basic understanding of the energy flows, chemical transformations, and the scale of the systems that support our current way of life. These concepts are fundamental to chemical engineering. Chemical engineers should be on the vanguard of separating hype from truly practical options yet occationally we let society down. This talk will present several examples where the fundamentals of chemical engineering can be used to access the practicality of proposed solutions to energy and water issues, as well as explain why the venture capital model struggles in the materials market.

352696 Hydrocarbons from Biogas Via Biogas Reforming, Syngas Cleaning and Catalytic Conversion

Abstract:

This research work focused on taking advantage of biogas as a cleaner, sustainable and renewable feedstock to produce liquid hydrocarbons. In this approach, biogas is first reformed to high quality synthesis gas (H_2 + CO) followed by Fischer-Tropsch synthesis (FTS). The Sustainable Energy Research Center (SERC) at Mississippi State University has developed and tested a continuous process over a pilot plant system to demonstrate the technology of biomass to liquid (BTL) fuels. The process includes biomass gasification, syngas conditioning and cleaning, and FTS steps. Currently, the typical gaseous product components from the gasifier are 20% H_2, 19% CO, 12% CO_2, and 47% N_2 if using wood chips as the feedstock. The N_2 and CO_2 contents are too high, the total syngas (H_2 + CO) content is only about 40%, and H_2/CO ratio of the bio-syngas is less than 1. The biomass-derived product gas is not ideal for the hydrocarbon synthesis. Exceptional product gas is produced from biogas (Table 1), it contains high concentrated syngas (CO+ H_2 > 90%), low impurities and CO_2, as well as N_2,H_2/CO ratio is between 1 and 2 which makes it perfect production of liquid hydrocarbon by FTS. The liquid products generated by FTS can be readily upgraded to transportation fuels since they are primarily composed of hydrocarbons initially.

Table 1 Product gas composition for simultaneous dry and partial oxidation (using air) of biogas.

Temperature (°C)	H_2 (%)	CO (%)	N_2 (%)	CH_4 (%)	CO_2 (%)	H_2/CO
750	48.96	40.82	6.33	1.75	2.01	1.2
775	49.45	41.22	6.27	1.3	1.74	1.2
800	50.06	42.2	6.05	0.96	1.12	1.19
825	50.57	42.7	6.03	0.57	0.78	1.18
850	50.75	43.11	5.93	0.4	0.25	1.18

351056 Research for an Energy Future

Abstract:

Global demand for electrical energy is increasing even as environmental concerns are changing the ways in which we produce it. Southern Company is a leading US producer of electricity and is dedicated to developing the latest technologies of pollution control, energy production, and energy efficiency, and bringing them to economic feasibility. This talk will highlight efforts in all these areas, including Southern Company's major research projects, research centers, and commercial operations. The talk will primarily focus on Southern Company's roles in developing pollution control technology, including acid gas scrubbers, selective catalytic reduction of NOx, and mercury control (including the Mercury Research Center and the High-Sulfur Pilot Baghouse Research Center). CO_2 capture will be discussed including a 25-MW demonstration plant and ongoing work at the National Carbon Capture Center, located at Alabama Power's Gaston Plant. Changing energy production methods including commercial operations of coal gasification, wind, and solar will also be discussed, as well as the new Solar Power Research Center. Efforts to develop technologies which conserve water in power production will be highlighted by a discussion of the Water Research Center located at Georgia Power's Bowen Plant. Finally, ongoing research to improve energy efficiency for our customers will be highlighted, as well as electric transportation research which would have major environmental benefits.

Sustainability and Environmental Impacts of Biofuel

Shijie Liu

Department of Paper and Bioprocess Engineering
SUNY College of Environmental Science and Forestry

1

Imagine a renewable world

2

Outline

- **The Hype of Biomass**
- **What is Sustainability?**
- **Biomass Crop**
- **Ethanol: Good or Bad**
- **Ethanol: a Platform chemical**
- **Ethanol: Biomass conversion efficiency**
- **Conclusion**

The Hype in Biomass

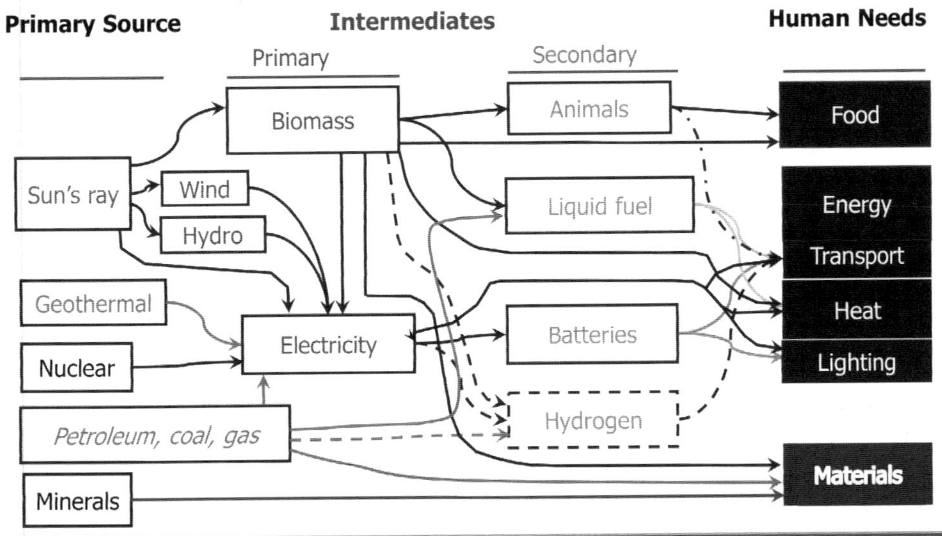

Where do we get our chemicals today?

- Petroleum, coal and natural gas
  - Dominant source of energy for human use
  - Materials use - plastics
- Biomass
  - Food
  - Heating use on the steep declining slope
  - Materials – dwellings, furniture
- Solar energy
  - Limited use in heating and lighting

5

What is Renewability?

- Merriam-Webster:
  - 1. Renewable – capable of being renewed
  - 2. Renewable – capable of being replaced by natural ecological cycles or sound management practices
- Point on a complete cycle

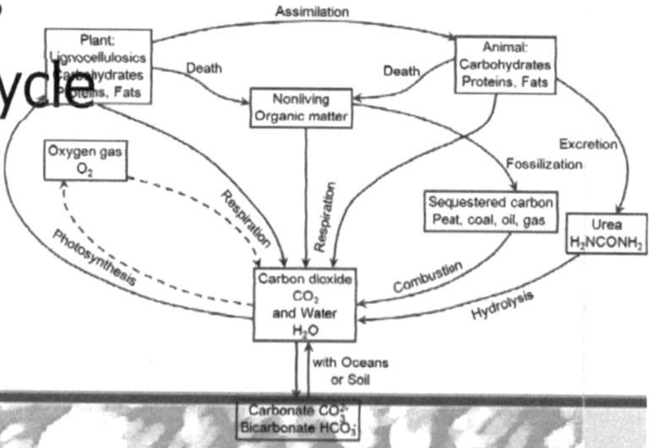

What is Sustainability?

- Merriam-Webster:
  - 1. Sustainable – capable of being sustained
  - 2. Sustainable – of, relating to, or being a method of harvesting or using a resource so that the resource is not depleted or permanently damaged.
- Non-trivial Steady State

(S. Liu, Bioprocess Engineering: Kinetics, Biosystems, Sustainability, and Reactor Design, Elsevier, 2012)

Biomass

- Agricultural crop
  - Cereal grain; Tubers
  - Straws
  - Grass
- Forest biomass
  - Trees
- Long rotation crops
  - Fruits, nuts
  - "Short-rotation" woody biomass

Is Forest Biomass Renewable?

- Manomet Center for Conservation Sciences report "Biomass Sustainability and Carbon Policy Study" also know as the Massachusetts Biomass Report (June 2010)
  - Woody biomass utilization sending more carbon into the atmosphere than fossil
  - Equal setting of forest biomass use with fossil use will occur after 32 years IF ...
  - Forest biomass carbon neutrality will not occur after 90 years

9

Massachusetts' report

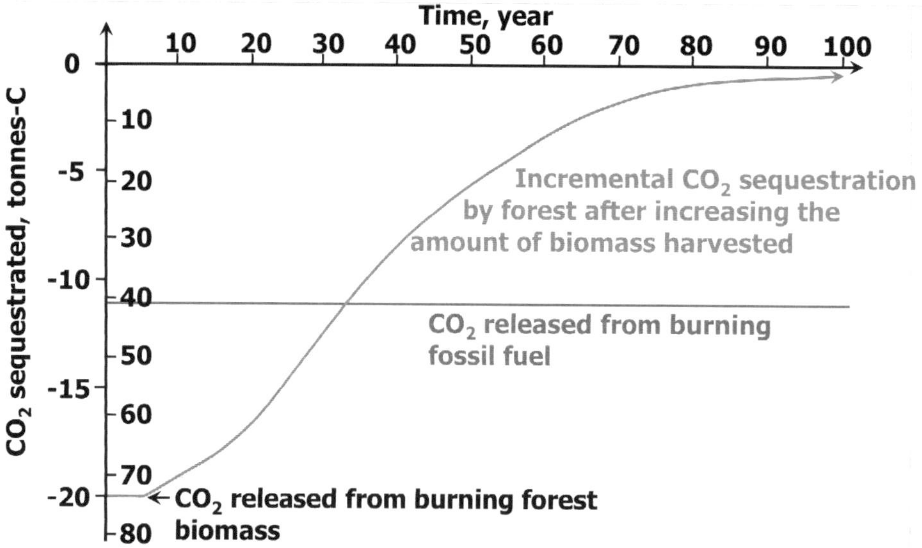

10

Biomass: Good or Bad?

- Fossil source
  - Energy dense: low CO_2 emission
- Biomass, CO_2 is CO_2
  - Less energy content (already oxygenated) - bulky
  - Not accounting: Forest disturbance – only half storage grows back when converting an untouched forest to a managed forest and equal amount underground in jeopardy
  - More CO_2 in atmosphere at sustainability
  - Hydrocarbon emission, e.g. isosprene

Sustainability – Biomass Production

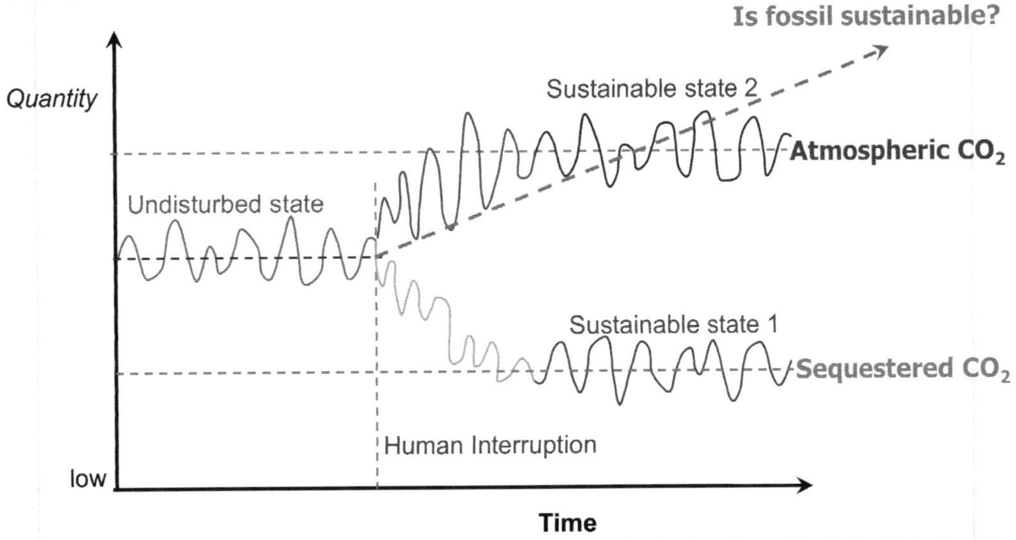

What is the Magic Biomass (Crop)?

Table 1. Recycle times and productivities of chemical/energy feedstock (Liu, 2010).

Feed stock	Recycle time	Standing Biomass tons/ha	Biomass Production tons/(ha·year)
Algae	1 month	0.9	11.25
Agricultural crops	3 month - 1 year	4.5	2.93
Temperate grasses	1 year	7.2	2.70
Savanna	1 year	18.0	4.05
Shrubs	1 – 5 years	27.0	3.15
Tropical forest	5 – 25 years	202.5	9.90
Trop. season forest	5 – 25 years	157.5	7.20
Boreal forest	25 – 80 years	90.0	3.60
Temperate deciduous	10 – 50 years	135.0	5.40
Temperate evergreen	10 – 80 years	157.5	5.85
Oil, gas and coal	280 million years	(38.4×10^{21} J)	(0)

Standing Biomass (~ X_∞)

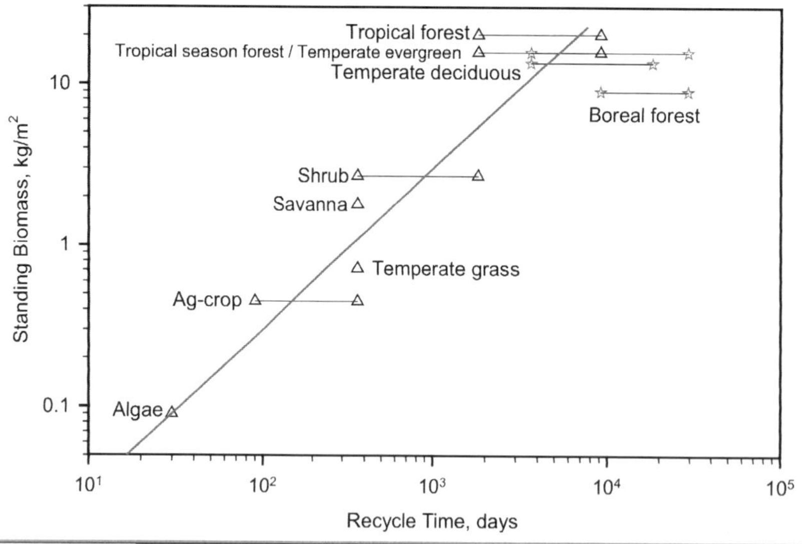

14

Biomass crop selection

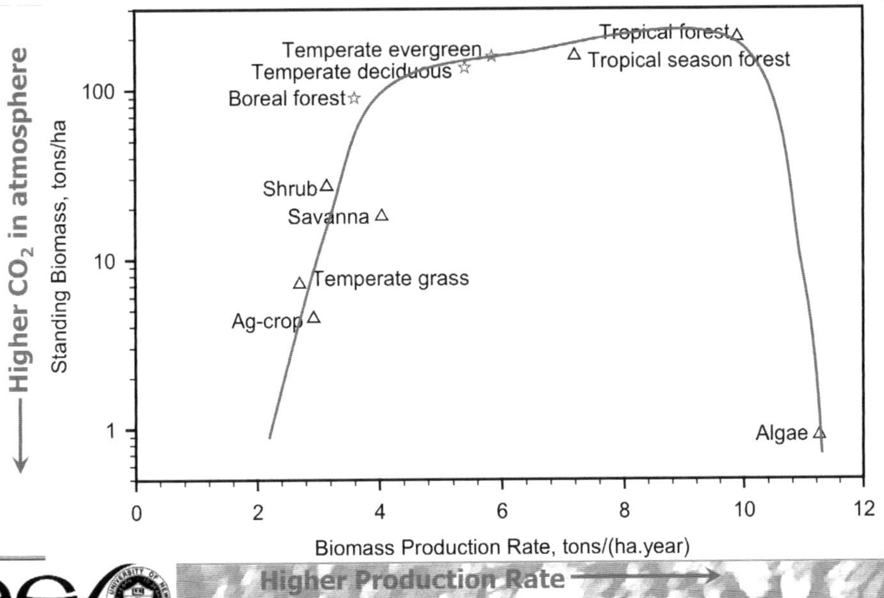

Biomass Crop Selection

- Productivity: rate of supply
- Standing biomass: CO_2 sequestration
- Water, Soil condition, Terrain
- Neighboring community
- Food needs, why "corn" is a better biomass
- Production cost
- There is no magic crop

Integrated Biorefinery:

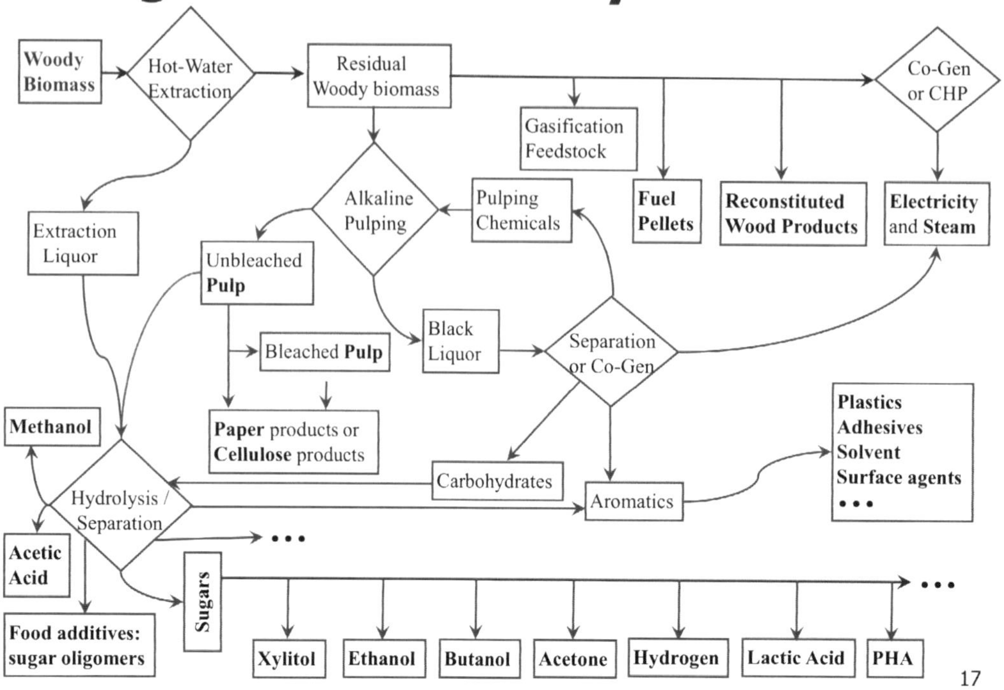

17

Ethanol: Good or Bad

- Gasoline additive: E10, going for E15
  - Commercial
  - Infrastructure established, and production facility design mature for new ones
  - Blending at distribution centers
  - Transportation
- Cereal grain use
- Physical and chemical properties

Ethanol: Platform chemical

- Jet fuel
- Bioplastics or bioploymers
- Hydrocarbons

$$CH_3CH_2OH \longrightarrow CH_2{=}CH_2 + H_2O$$

$$2CH_3CH_2OH \longrightarrow CH_2{=}CH\text{-}CH{=}CH_2 + H_2O + 2H_2$$

Wang, Y. and Liu, S. J. Bioproc. Eng. Biorefinery, 1(1), 33-43.

Ethanol: Biomass conversion

- Theoretical yield: 51% on sugars
  - Anaerobic fermentation, can tolerate O_2
  - Fermentation to high concentrations
- As compared with
- Butanol – low yield, low concentration
- PHA – low yield, high recovery cost

Fermentation: Wood extract Hydrolysate

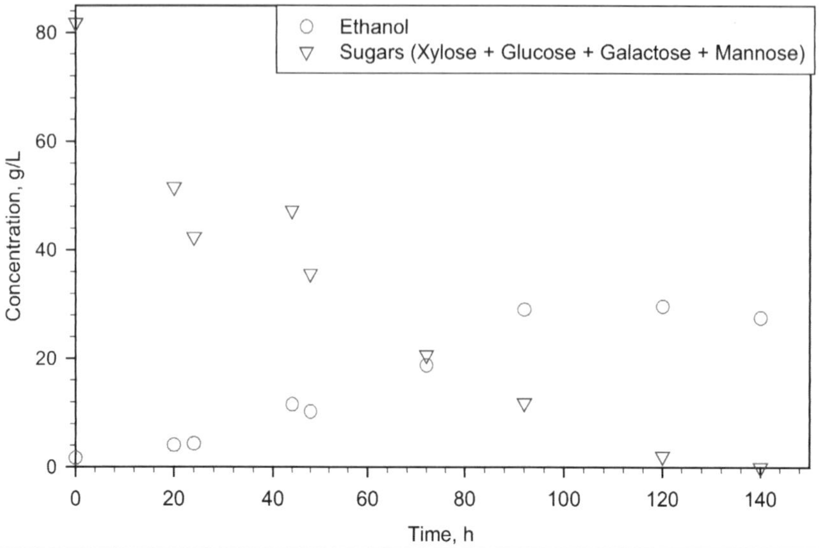

Butanol fermentation

Glucose in feed: 45 g/L

Acetone	5.2	g/L
Ethanol	1.8	g/L
Butanol	9.8	g/L
Total	16.8	g/L

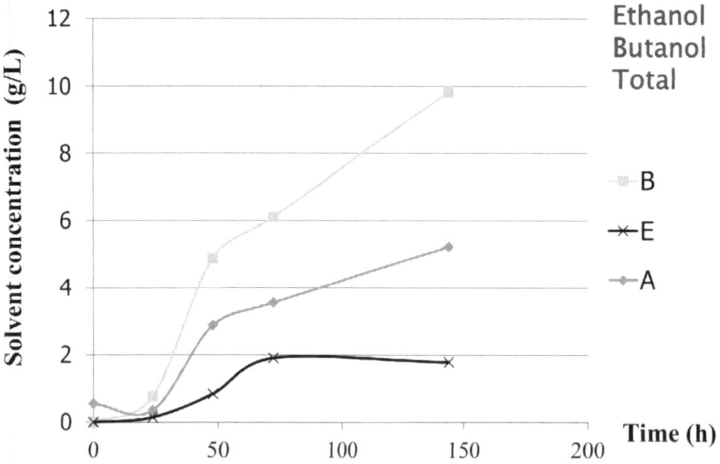

Butanol fermentation

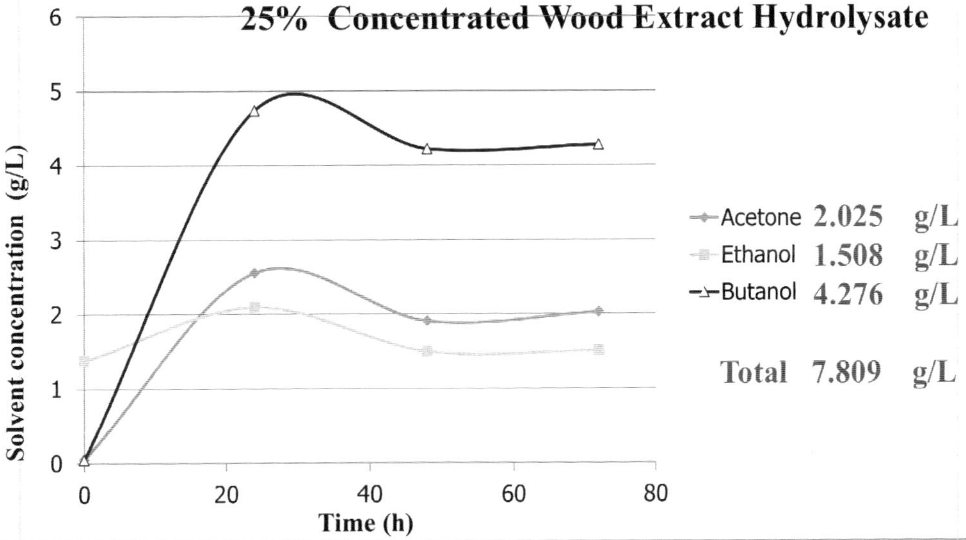

Conclusions

- Future depends on biomass, despite a hard choice to make: CO_2 is CO_2
- Biomass is both renewable and sustainable
- Ethanol is a good choice for continuing developments in renewable chemical and energy sectors
- Ethanol is a platform chemical, from which products know today from petroleum can be produced

24

Thanks!

25

PROSYS™ Maximize Your Plant.

Operator Errors and What Can be Done to Minimize

Global Congress on Process Safety

March 31, 2014

10TH GLOBAL CONGRESS ON PROCESS SAFETY

PROSYS™ Maximize Your Plant™

10TH GLOBAL CONGRESS ON PROCESS SAFETY

Presenter

Tom Nolan

- Graduated from Ohio University with B.S. in Chemical Engineering
- 24 years experience in chemical process industry in a variety of roles
- MAIC and DFSS Six Sigma Black Belt

PROSYS Maximize Your Plant™ 10th Global Congress on Process Safety

The Cost of Errors

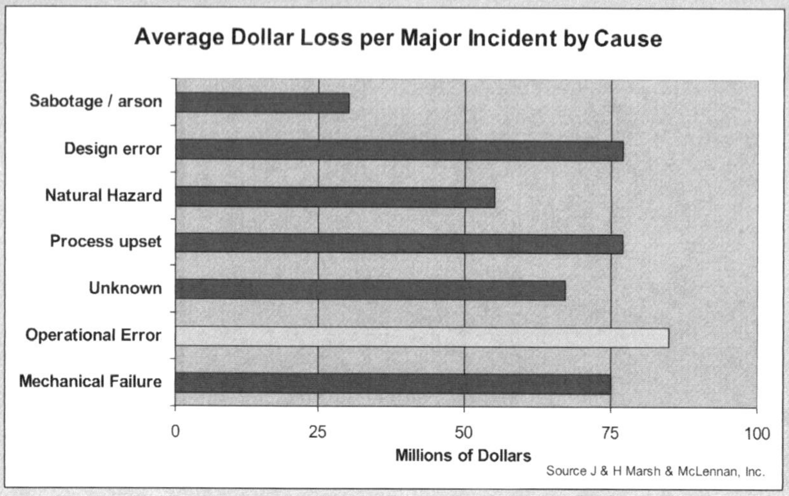

PROSYS Maximize Your Plant™ 10th Global Congress on Process Safety

The Cost of Operator Errors

- **ASM estimates total loss due to operator error is $8B per year**
- **Chemical Safety Topical Committee - average of one chemical incident per day - cost of over $2 million per incident to comply with requirements**
- **Errors cause 42% of unscheduled shutdowns**
- **70% of process incidents occur during start-up or shutdown**

PROSYS Maximize Your Plant™

10TH GLOBAL CONGRESS ON PROCESS SAFETY

Reasons for Errors

- Lack of Skill
- Lack of Knowledge
- Carelessness
- System Design
    - Operator set up to make errors by inappropriate design – built in errors
    - Errors that are predictable are preventable by better design

PROSYS Maximize Your Plant™

10TH GLOBAL CONGRESS ON PROCESS SAFETY

Focus of this Discussion

- System Design
    - **Alarm Management**
    - **Operator Graphics**
    - Operator set up to make errors by inappropriate design – built in errors
    - Errors that are predictable are preventable by better design

PROSYS Maximize Your Plant™

10th GLOBAL CONGRESS ON PROCESS SAFETY

Alarm Floods

In a number of industrial incidents, alarm floods were identified as a significant contributing cause to the incident…

As found by EEMUA in 1999 and CSB

Alarm Flood defined by ISA 18.2 as -
"10 or more annunciated alarms in any 10 minute period per operator"

PROSYS Maximize Your Plant™

10th GLOBAL CONGRESS ON PROCESS SAFETY

Why do Alarm Floods Occur?

One reason is not providing dynamic alarm management

- **Alarms need to indicate abnormal situations that require operator action**
- **Processes do not operate in one state**
- **What is normal vs abnormal changes with state**
- **Alarms are typically configured for run therefore many alarms are triggered upon a Change of State: Run to Shutdown**
- **Many of the alarms are not applicable or actionable for the new state impleading the operators ability to act quickly on what is important**

PROSYS Maximize Your Plant™

10TH GLOBAL CONGRESS ON PROCESS SAFETY

What is Impacted by Alarm Floods?

- **Product quality**
- **Operability or profitability of the process**
- **Loss of equipment**
- **Loss of containment – environmental releases**
- **Injury and loss of life in plant or community**

PROSYS Maximize Your Plant™

10TH GLOBAL CONGRESS ON PROCESS SAFETY

What Makes Alarm Floods so Dangerous?

Can be a problem for three reasons:

- **A deluge of alarms can cause critical alarms to be missed**
- **Floods can be a significant distraction when dealing with process upsets**
- **Can be an indicator of larger systemic safety issues**

PROSYS™ Maximize Your Plant™

10th GLOBAL CONGRESS on PROCESS SAFETY

Impacting Alarm Management Design

- **Alarm rationalization is <u>not</u> a process to eliminate alarms - its about quality**
- **Good rationalization will add alarms when appropriate**
- **Typical (Static) rationalization is only for run mode**
- **Dynamic rationalization considers all plant modes**

PROSYS™ Maximize Your Plant™

10th GLOBAL CONGRESS on PROCESS SAFETY

Dynamic Alarm Management

- **Dynamic rationalization does everything a static rationalization does plus asks "when" for each alarm**
- **Requires dynamic software to make changes based upon operating mode of the plant**
- **Eliminates redundant alarms and lowers operator loading during transitions**
- **Only alarms what is abnormal and actionable for the given state**

PROSYS Maximize Your Plant™

10TH GLOBAL CONGRESS ON PROCESS SAFETY

Actual Performance Metrics vs ISA 18.2

Type	Avg Alarm Rate/hr	Low/Hi of Avg Rate/hr	Peak Alarm Rate/hr	Low/Hi of Peak Rate/hr	% Time in Flood	Standing Alarms >24 hrs
Before Rationalization	30	Low=6.2 Hi=61	638	Low=152 Hi=2402	17.2%	9
ISA 18.2 Target Metrics	6	---	<60	---	<1%	<5
After Dynamic Alarm Management	2	Low=0.09 Hi=2.5	25	Low=13 Hi=42	0.25%	4

PROSYS Maximize Your Plant™

10TH GLOBAL CONGRESS ON PROCESS SAFETY

Results After Dynamic Alarm Management

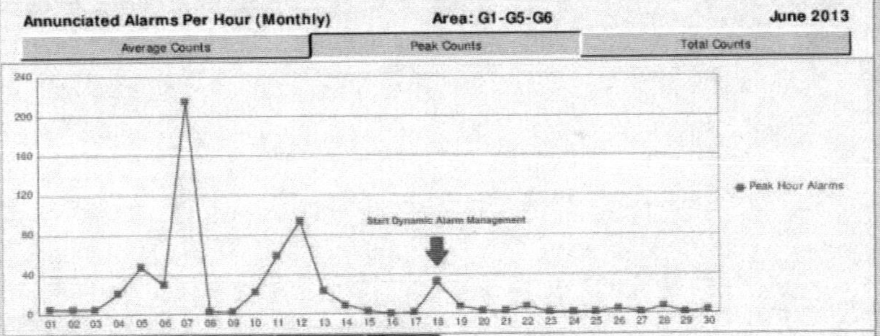

Day	Min Hour Alarms	Avg Hour Alarms	Peak Hour Alarms	Total Alarms
01-Jun-2013	0	0.3	5	6
02-Jun-2013	0	0.4	5	9
03-Jun-2013	0	0.5	6	13
04-Jun-2013	0	1.1	21	27
05-Jun-2013	0	13.3	48	319
06-Jun-2013	0	3.6	30	86
07-Jun-2013	0	11.9	216	285
08-Jun-2013	0	0.4	3	10
09-Jun-2013	0	0.3	2	7
10-Jun-2013	0	1.5	23	35
11-Jun-2013	0	4.7	59	112
12-Jun-2013	0	11.9	94	285
13-Jun-2013	0	3.8	24	90
14-Jun-2013	0	1.3	9	32
15-Jun-2013	0	0.3	3	8

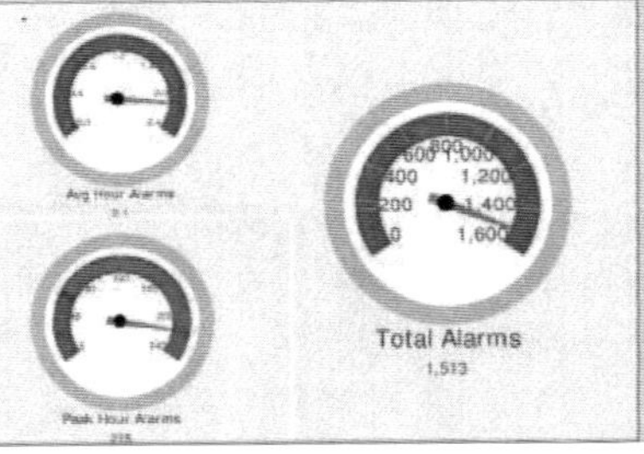

Operator Graphics – The Risk of Changing Focus

- Distraction of changing focus from process graphic to faceplate window
- Added workload from managing multiple open windows
- Increased probability of errors when changing values for un-intended tag with multiple faceplates open
- All of above become multiplied and more complex when process is transitioning from one state to another

Faceplate Operation

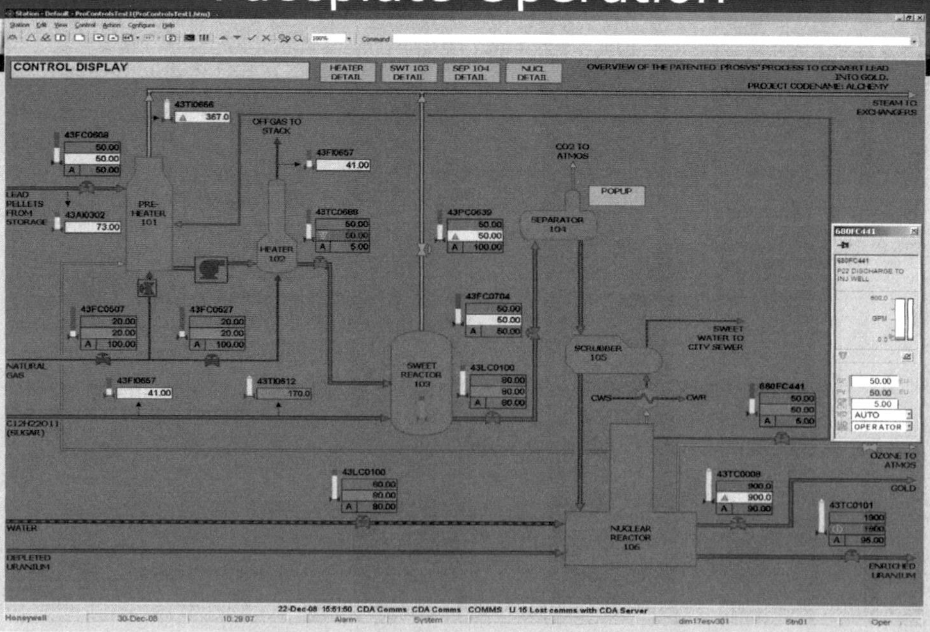

PROSYS Maximize Your Plant™

10th GLOBAL CONGRESS ON PROCESS SAFETY

Potential for Errors

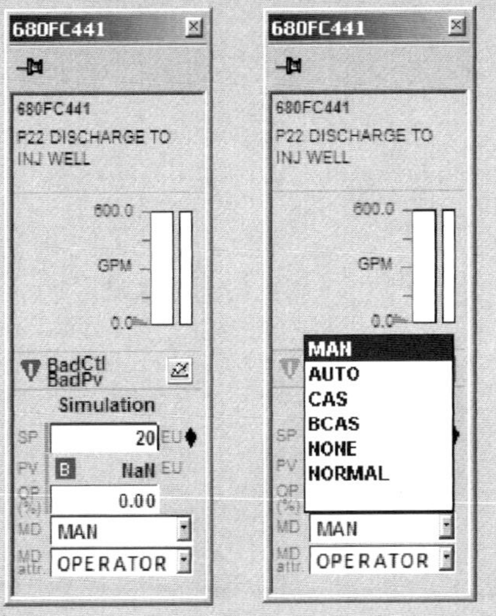

Faceplate Design -can introduce additional errors

- Mode drop down list covers SP, PV and OP values
- Operator may select wrong Mode if SP, PV and OP values are hidden from view
- Mode list offers more options than necessary

PROSYS Maximize Your Plant™

10th GLOBAL CONGRESS ON PROCESS SAFETY

Direct Entry Operation

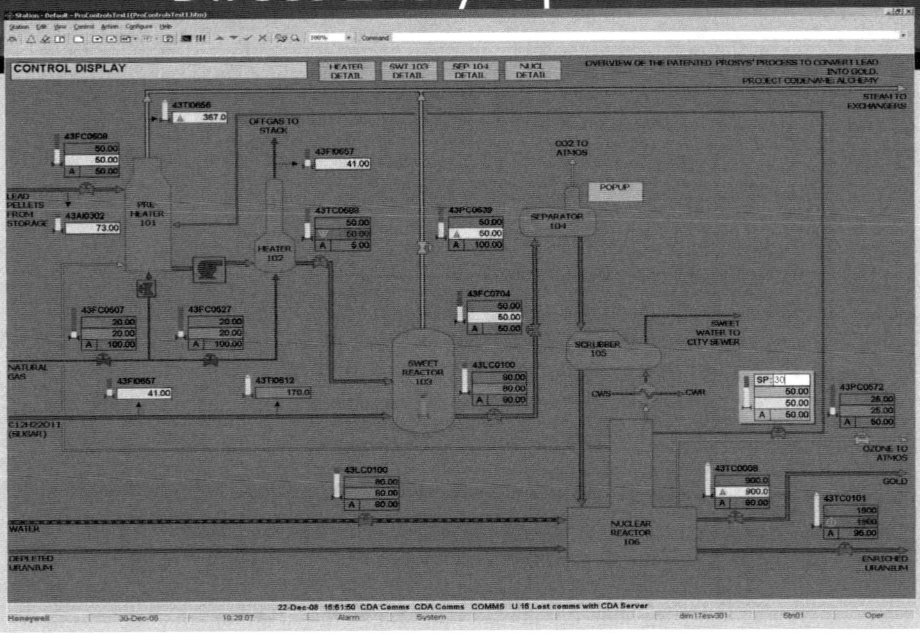

Direct Entry Fields

Values can be entered by:

- Select point
- Typing value via keypad
- Pressing enter

Key Factors:

- By maintaining focus of points after value change, it improves safety for input errors, multiple clicking functions and is more time efficient
- Easily recognizable operator enterable fields

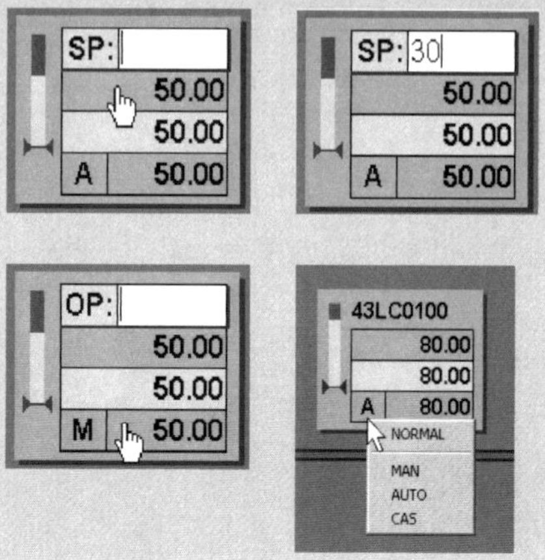

PROSYS Maximize Your Plant™
10TH GLOBAL CONGRESS ON PROCESS SAFETY

Evaluation of Methods

Comparison of Faceplate to Direct Entry

- **Keystroke Level Analysis**
- **Operator Loading Analysis**
- **Risk Analysis**

Keystroke Level Model

Symbol	Time (s)	Description
K	0.28	Keystroke
P	1.1	Point to object
BB	0.2	Click on object
H	0.4	Home hands on keyboard or mouse
M	1.2+	Mental act or routine thinking

PROSYS Maximize Your Plant™

10th GLOBAL CONGRESS ON PROCESS SAFETY

Estimate Execution Time Analysis

- **Evaluates execution time by an operator comparing faceplates to direct entry for SP/OP and Mode Changes**

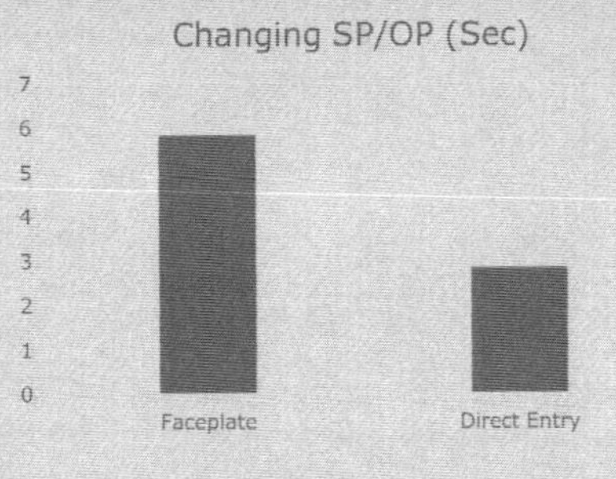

- **50 % Reduction**

PROSYS Maximize Your Plant™

Estimated Operator Loading Analysis

- **Evaluates Mental and Physical Load on Operator**
  - Tasks that require very little thought are removed – such as (BB), (H)

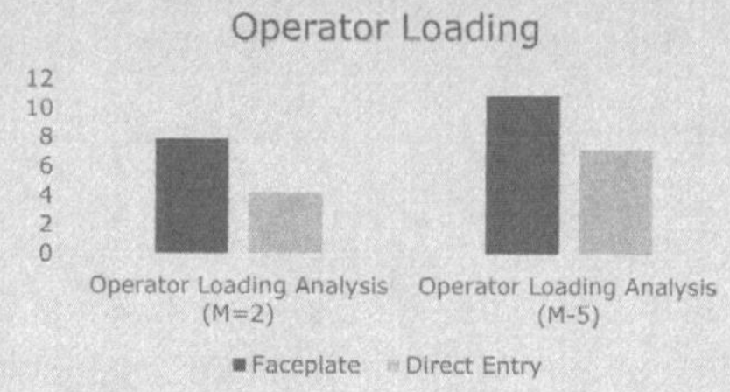

  - Mental and Physical tasks are weighted based on the amount of fatigue they cause – assigned a weight of M=2.0 to 5.0

PROSYS Maximize Your Plant™

Risk Analysis

- **Identifies actions where operator entry errors can occur**

PROSYS™ Maximize Your Plant™

10TH GLOBAL CONGRESS ON PROCESS SAFETY

Risk Analysis

Faceplate operation KLM for changing SP/OP	Potential Error
Move to shape (P)	Select wrong point
Click on shape (BB)	
Move to faceplate (P)	Point in faceplate is previous point
Click on SP/OP field (BB)	Change wrong parameter
Move hands to keyboard (H)	
Type in value and press enter (4K)	Mistype value and press enter
Move hands to mouse (H)	
Move to close faceplate (P)	
Click to close faceplate (BB)	
Potential error count	**4**

PROSYS™ Maximize Your Plant™

10TH GLOBAL CONGRESS ON PROCESS SAFETY

Risk Analysis

Direct Entry operation KLM for changing SP/OP	Potential Error
Move to SP/OP (P)	Select wrong parameter or point
Click on SP/OP field (BB)	
Move hands to keyboard (H)	
Type in value and press enter (4K)	Mistype value and press enter
Potential error count	**2**

Direct Entry Vs Faceplate

Summary of Benefits

- **Execution Time SP/OP Changes - 51.5% less**
- **Execution Time for Mode Changes – 50% less**
- **Operator Loading – 33 % to 47% less**
- **Risk of Errors – 50% less**

Conclusion

- **Many industrial errors are a result of operators using control systems with flawed designs**
  - Poor design and performance by alarm management distracts operators and/or occludes critical alarms
  - Operator graphics with multiple open faceplates can cause a change intended for one controller to be entered into the faceplate of another
- **The two mechanisms listed above are predictable and therefore preventable through better design available today**

PROSYS™ Maximize Your Plant.

Questions?

Comments?

10TH GLOBAL CONGRESS ON PROCESS SAFETY

Problem Statement

visium KMS
Process Safety Management Solution

- Any organisation can improve by leveraging knowledge and lessons learned.
- Knowledge can be externally or hidden locally, in a global enterprise, where lessons learned are difficult to share beyond their initial impact.
- There is so much data, information and knowledge (opportunities) coming at your company that can make you stronger or make you look like your not paying attention.
- Not paying attention or managing the available knowledge, can sink your ship… nobody wants that!

Expert Insight Delivered

2

Osys
Part of the Rolls-Royce Group

What is Knowledge Management?

visium™KMS
Process Safety Management Solution

- Knowledge management is the concept in which an enterprise consciously and comprehensively gathers, organizes, shares, and analyses its knowledge to further its aims.
- Turning data into information is nice.
- Turning Information into Knowledge is good!
- Turning Knowledge into Wisdom is great!
- But something is missing…
- The CAPA (doing) loop…

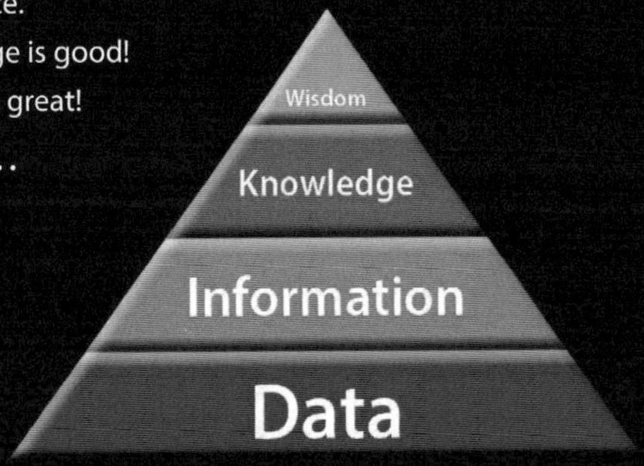

Part of the Rolls-Royce Group

What we need is this…The HUAA MS

visium™KMS
Process Safety Management Solution

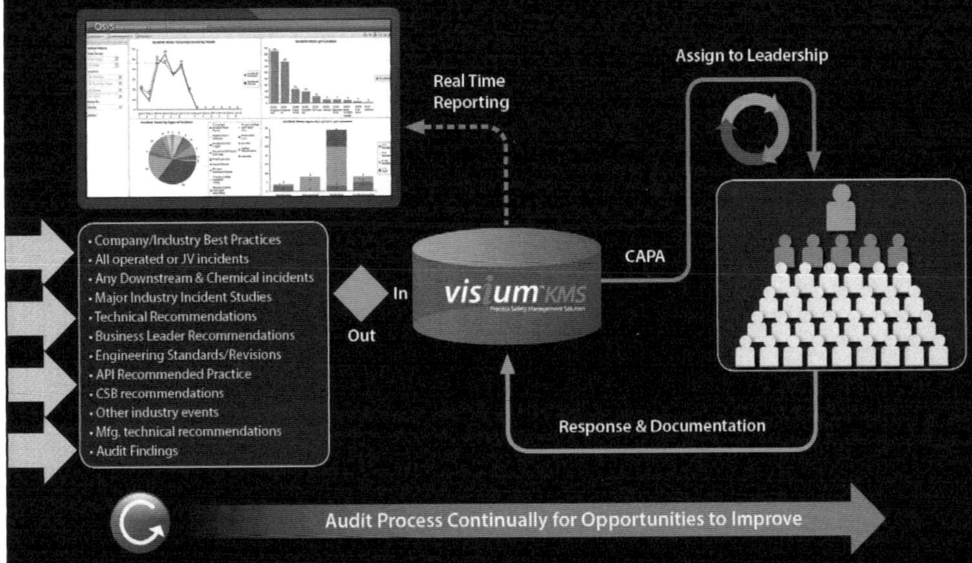

Part of the Rolls-Royce Group

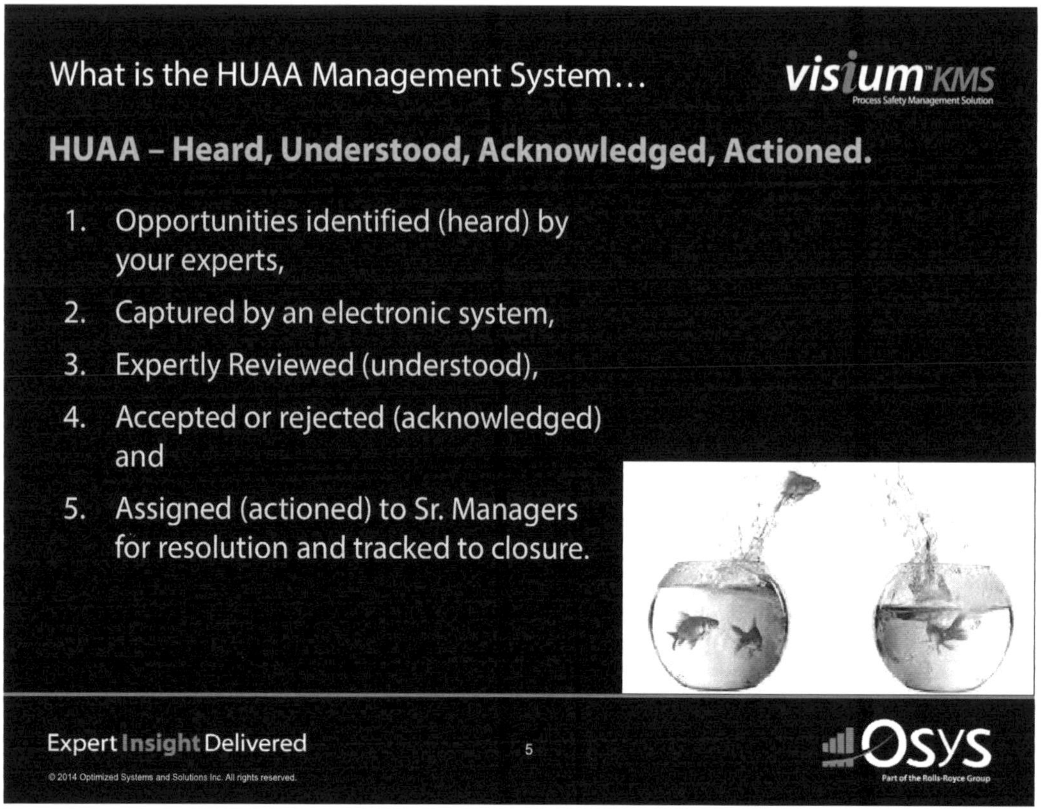
What is the HUAA Management System…
visium KMS
Process Safety Management Solution
HUAA – Heard, Understood, Acknowledged, Actioned.
1. Opportunities identified (heard) by your experts,
2. Captured by an electronic system,
3. Expertly Reviewed (understood),
4. Accepted or rejected (acknowledged) and
5. Assigned (actioned) to Sr. Managers for resolution and tracked to closure.
Expert Insight Delivered
5
© 2014 Optimized Systems and Solutions Inc. All rights reserved.
OSYS
Part of the Rolls-Royce Group

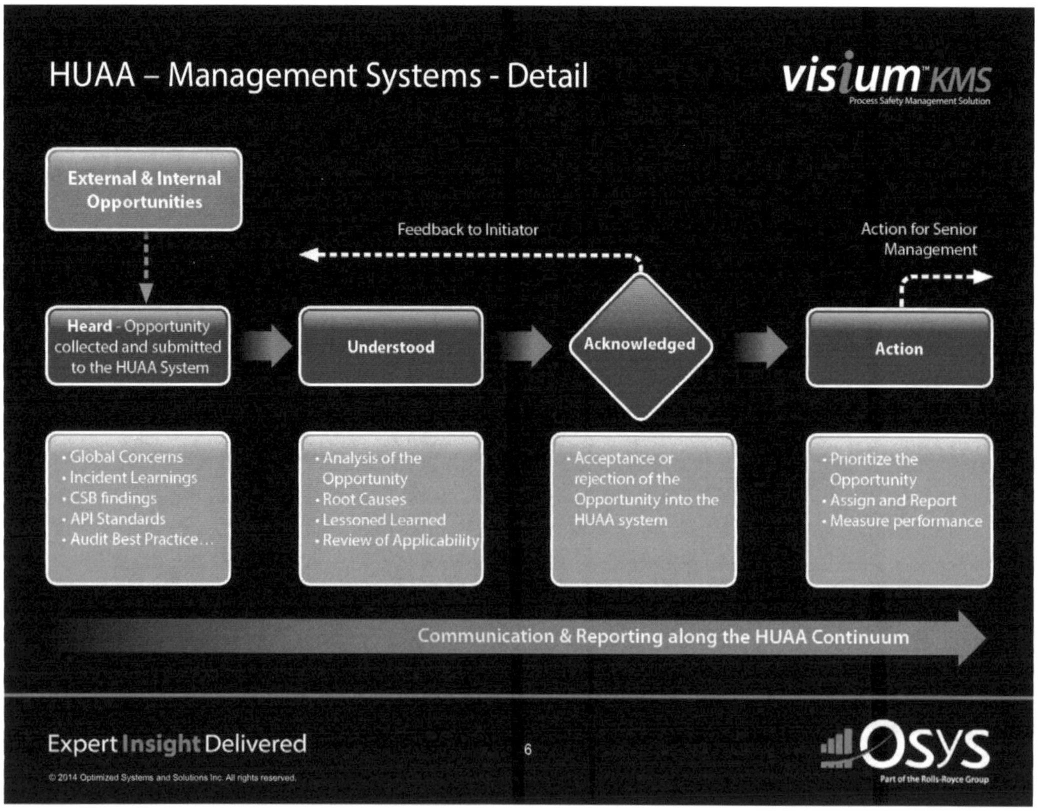
HUAA – Management Systems - Detail
visium KMS
Process Safety Management Solution
External & Internal Opportunities
Feedback to Initiator
Action for Senior Management
Heard - Opportunity collected and submitted to the HUAA System
Understood
Acknowledged
Action
• Global Concerns
• Incident Learnings
• CSB findings
• API Standards
• Audit Best Practice…
• Analysis of the Opportunity
• Root Causes
• Lessoned Learned
• Review of Applicability
• Acceptance or rejection of the Opportunity into the HUAA system
• Prioritize the Opportunity
• Assign and Report
• Measure performance
Communication & Reporting along the HUAA Continuum
Expert Insight Delivered
6
© 2014 Optimized Systems and Solutions Inc. All rights reserved.
OSYS
Part of the Rolls-Royce Group

HUAA Results

- Efficient Collection - Important Opportunities are identified and collected from many sources (internal and external), and entered into the HUAA management system.
- Expert Vetting - These opportunities are vetted, parsed and resolved for inclusion or exclusion; documenting & communicating both.
- Accountable Assignment - Included opportunities are systematically assigned to Sr. Leaders and managed across the enterprise at SVP and C levels.
- Action - Improvement will be assured through the accountable actions of these leaders.
- Results - The organization will gain competitive advantage by active listening, learning from their experience, as well as the experience of others, and by taking ACTION!

Expert Insight Delivered 7

Sharing is not enough to keep your ship afloat!

Expert Insight Delivered 8

352631 Autocad - Smart Grid – Agriculture - Architecture and the United States Government

Abstract:

The computational methods, process control, process design, process safety, public policy of the smart chip for agriculture industry, and architecture industry in the United States utilizing the computational methods of AutoCAD in the structural development stages of the smart grid for device/component/equipment and geographical development and public policy for grant funding.

352747 Production, Characterization and Catalytic Studies of Biobased Carbon Materials

Abstract:

Carbon-based nano materials were produced from wood chips, lignin and bio-char. The influence of reaction conditions on the production of nano-carbons was investigated. Carbon-based nanomaterials were characterized using scanning electron microscopy (SEM), transmission electron microscopy (TEM), high resolution transmission electron microscopy (HRTEM), and X-ray diffraction (XRD). The catalytic performance of these nanomaterials as syngas cleaning adsorbents and chemical conversion catalysts for bio-fuel processes has been evaluated.

347835 Uncertainty in Sour Gas Viscosities Estimation, What Is the Effect on Your Reservoir Inflow and Tubing Performance

Abstract:

In light of the need to find an applicable, consistent and promising correlation to accurately calculating sour gases viscosity under several reservoir conditions, this study aims to compare the existing correlations and other industry standard methods, check their accuracy and range of applicability for sour/acidic gases. Moreover, investigating the impact of uncertainties in estimating sour gas viscosities on reservoir flow performance relationship (IPR), reserves and resources estimations, tubing performance, pipeline and facilities design, the design of compressors and more importantly the need to find such correlation to be used in further advancements of reservoir simulation.

354052 Chromonic Nanocarriers for Chemotherapeutics: Size Distribution and Control Release Studies

Abstract:

Conventional chemotherapy fails to release the drug in the controlled manner; rather they release the drug instantly due to which re-administration is required as therapeutic index falls. Moreover, they target both the cancer cells and non-cancer cells which make the treatment of the cancer cells highly ineffective due to excessive toxicities. Recently, the nanoparticle mediated drug delivery for chemotherapeutics has gained much attention. The present study reports the use of a novel class of materials – *Chromonics* in cancer drug delivery. Folic acid as a chromonic material has been explored to develop nanocarrier for sustained release of chemotherapeutic drugs *in-vitro*. With the help of a biocompatible polymer as a tool, folate nano-domains were formed from self-assembled liquid crystalline folate solutions. These nano-domains were crosslinked to form a stable nanoparticle system. The effect of different factors like crosslinking agent, relative polymer and folate concentration, on the size distribution was investigated in detail. Based on the observations, we have reported that not only the synthesis but also the size of the developed nanoparticles can be controlled by varying the folate: HPMC weight ratios. The present study then explored the release of drugs from these nanocarriers. We have investigated the release of folic acid as a model drug from these nanoparticles as well as release of methotrexate and cytarabine as an anticancer drug for the period of 60 days. It was observed that release of drug was nanoparticle's size-dependent, release medium dependent, pH of the release medium dependent, crosslinking agent and its concentration dependent. The significant part of the current study is the synthesis of chromonic nanoparticles, how the size can be controlled, how controlled release can be obtained as well as how the release profiles can be modified depending on the dosage forms. Therefore, we can say that folate nanoparticles can be a potent nanocarrier for the sustained release of cancer drugs.

Double Containment Piping Systems: A Fail-Safe, Leak-Free Solution

Darin Johnson, P.Eng.
Applications Engineer
IPEX USA LLC

AIChE – 2014 Spring Meeting
New Orleans, LA

Monday, March 31, 2014

Why Double Containment Piping Systems?

- December 1988 – law change within the ***Resource, Conversation, and Recovery Act*** in the United States regarding underground piping.
- U.S. Environmental Protection Agency (EPA) requirement: Code of Federal Regulations (CFR), Title 40, Part 280.
  - Petroleum Products
  - Hazardous Substances (over 3200, CFR Title 40 Part 302.4)
  - Hazardous Wastes

Why Double Containment Piping Systems?

- Buried tank - all buried piping connected to it must be double contained.
- Above ground tank – associated buried piping consists of 10% or more of the tank volume, it also must be double contained.

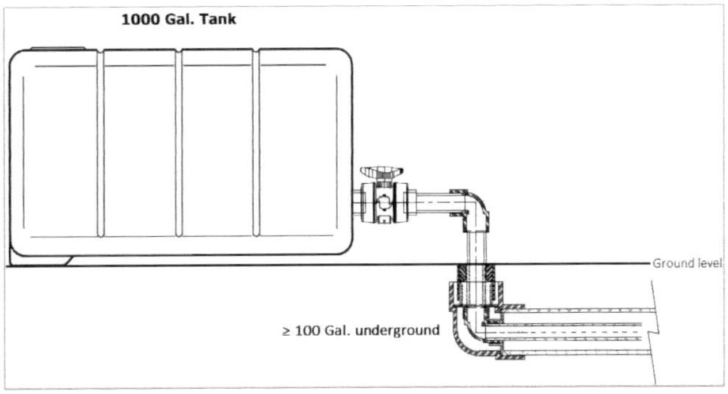

Why Double Containment Piping Systems?

- If the piping is less than 10% of the volume of the aboveground tank, and you choose not to double contain the buried piping, ***responsibility for any clean-up or remediation resulting from a release will still fall on the Client.***

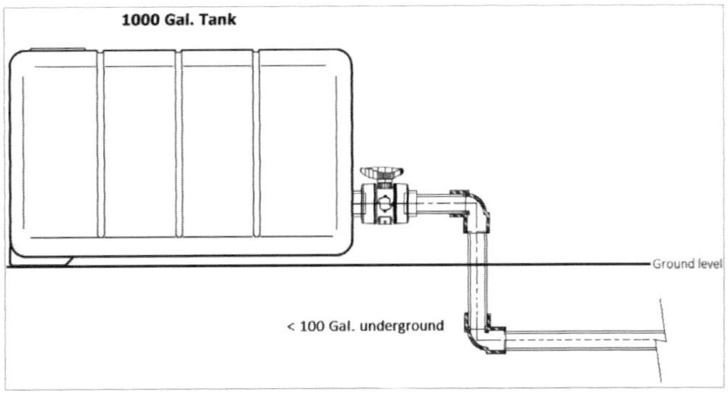

Why Double Containment Piping Systems?

- **Above Ground Piping Systems**
  - Different jurisdictions have to ability to require chemicals be double contained.

Nomenclature of Double Containment Piping

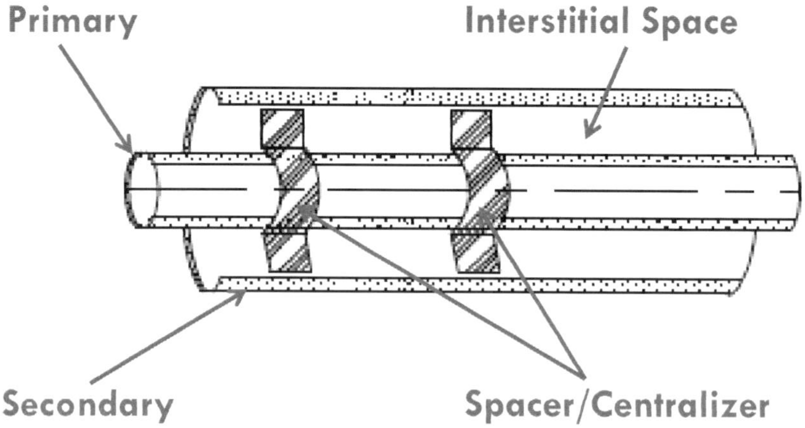

Double Containment Piping Systems: Material Combinations

SYSTEM Carrier	SYSTEM Containment	Carrier SIZE inches
PVC 80	PVC 80	1/2 - 12
PVC 80	PVC 40	1/2 - 12
PVC 40	PVC 40	1/2 - 12
PVC DWV	PVC DWV	1-1/2 - 12
PVC 80	CS SCH 10	1/2 - 12
CPVC 80	CPVC 80	1/2 - 8
CPVC 80	PVC 80	1/2 - 12
CPVC 80	PVC 40	1/2 - 12
CPVC 80	CS SCH 10	1/2 - 12
NAT PP 80	PVC 80	1/2 - 2
NAT PP 80	PVC 40	1/2 - 2
RED PVDF 80	CPVC 80	1/2 - 2
RED PVDF 80	PVC 80	1/2 - 2
NAT PVDF 80	PVC 80	1/2 - 2
ENFUSION PP	PVC DWV	1/2 - 6
S40 A53B CS BW	PVC 80	1/2 - 8
S40 A106B CS BW	PVC 80	1/2 - 8

SYSTEM Carrier	SYSTEM Containment	Carrier SIZE inches
S40 A53 BLK SW	FRP	1/2 - 2
S40 A53 BLK SW	PVC 80	1/2 - 2
S80 A53B CS SW	PVC 80	1/2 - 2
S10 T304L SS BW	PVC 80	1/2 - 8
S10 T304L SS BW	PVC 40	1/2 - 8
S10 T0304L SS BW	CPVC 80	1/2 - 8
S10 T304L SS BW	FRP	1/2 - 12
S10 T304L SS BW	S10 T304L	1/2 - 12
S40 T316L SS BW	PVC 80	1/2 - 8
S40 T316L SS BW	CPVC 80	1/2 - 8
S40 T316L SS BW	FRP	1/2 - 12
S40 T304L SS SW	PVC 80	1/2 - 2
S40 T304L SS SW	PVC 40	1/2 - 2
S40 T304L SS SW	CPVC 80	1/2 - 2
S40 T304L SS SW	FRP	1/2 - 2
S10 T316L SS SW	PVC 80	1/2 - 2
S10 T316L SS SW	CPVC 80	1/2 - 2

Double Containment Piping Systems: Design Considerations

- **Material Selection**
  - Primary Pipe
    - Chemical Compatibility
    - Pressure Capabilities
    - Thermal Properties
  - Secondary Pipe
    - Chemical Compatibility
    - Loading (Static & Live) for buried applications
    - Hanging requirements for above ground
    - Leak detection requirements
    - UV requirements

Double Containment Piping Systems: Design Considerations

- **Thermal Expansion & Contraction**
  - Two basic thermal relationships may be present:
    - When only one of the two systems experiences thermal expansion/contraction.
    - When both systems experience different magnitudes of thermal expansion/contraction.
  - The Floating System
  - The Restrained System

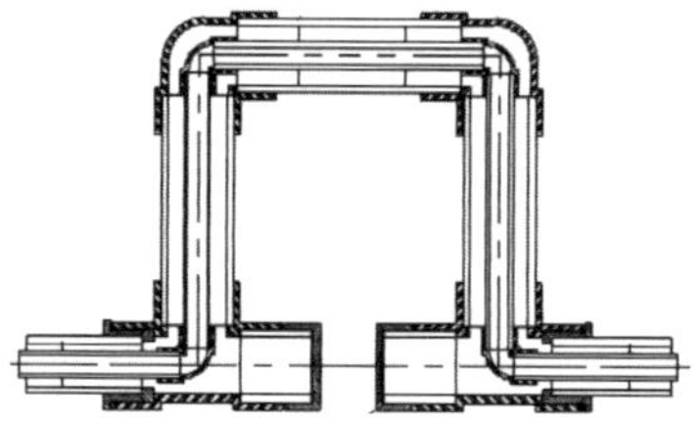

Double Containment Piping Systems: Leak Detection Requirements

- EPA requirement: (CFR), Title 40, Part 280
  - Pressurized systems require electronic leak detection that must be checked once per year.
  - Drainage systems (gravity flow) require visual leak detection with monthly manual inspection for a media leak.

Double Containment Piping Systems: Leak Detection Requirements

- The following forms of leak detection are acceptable as per regulations
  - Visual (drainage applications only)
  - Electronic low point probes
  - Continuous cable

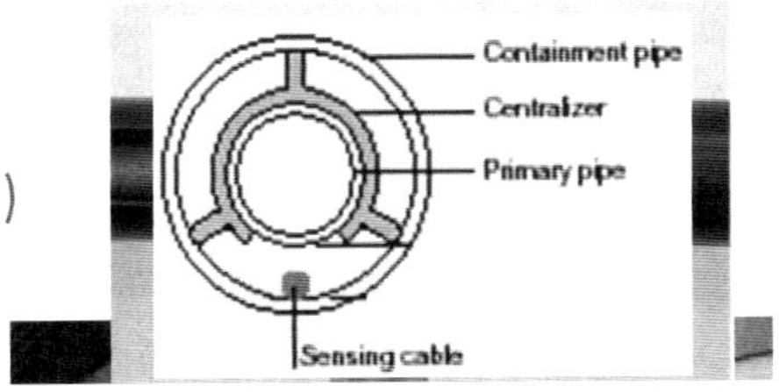

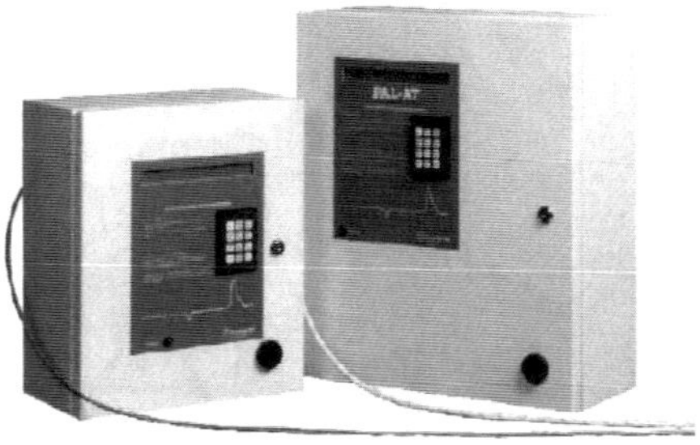

Double Containment Piping Systems: Applications

- Underground Applications
- WTP & WWTP
- Oil & Gas
- Hospitals
- Museums
- Food and Beverage
- Pharmaceuticals
- Laboratories
- Data Centers

Questions?

Thank you!

DTP Process

"Now Ready to License"

JGC Corporation
Mitsuo Morita

2014

JGC CORPORATION

MITSUBISHI CHEMICAL

What is DTP Process ?

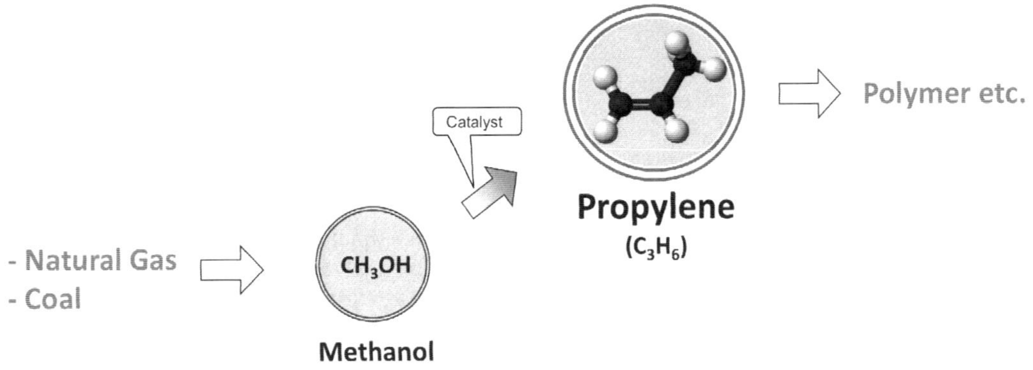

-On-Purpose Propylene Production from Methanol

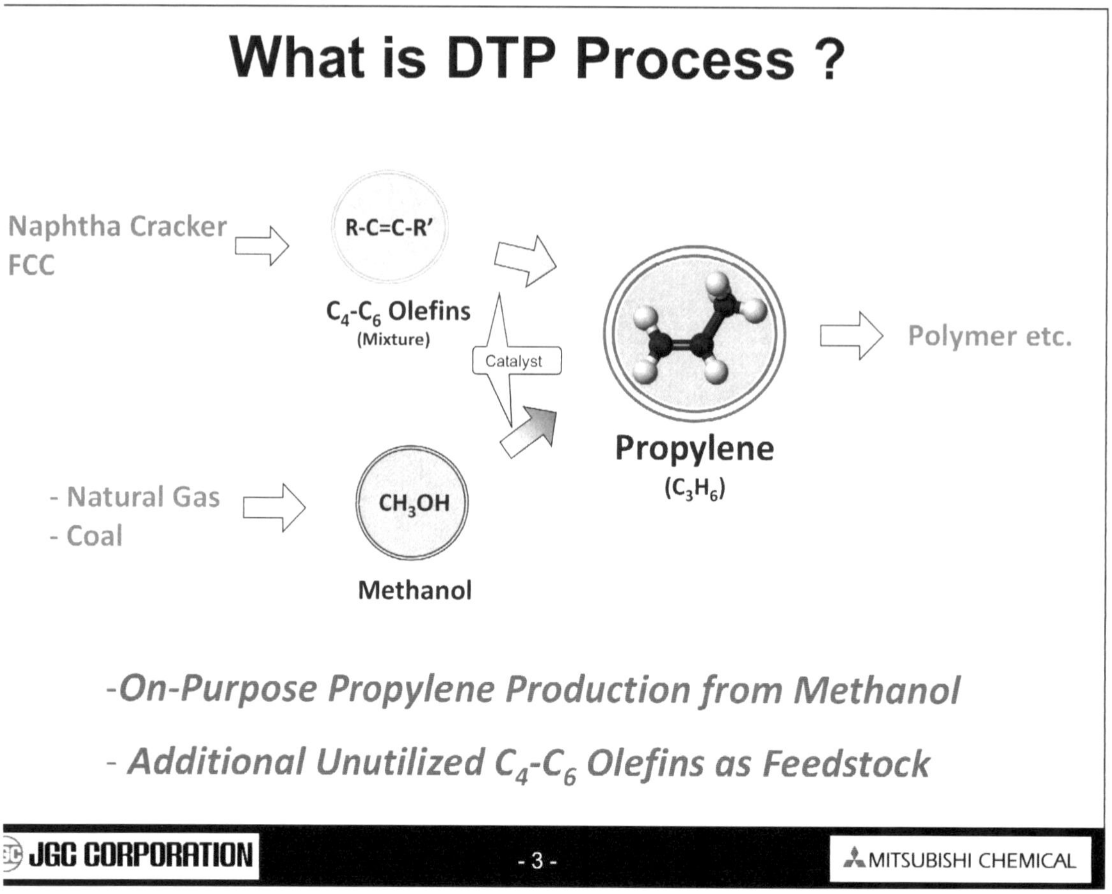
What is DTP Process ?
Naphtha Cracker
FCC
R-C=C-R'
C_4-C_6 Olefins
(Mixture)
Catalyst
Propylene
(C_3H_6)
Polymer etc.
- Natural Gas
- Coal
CH_3OH
Methanol
-On-Purpose Propylene Production from Methanol
- Additional Unutilized C_4-C_6 Olefins as Feedstock
JGC CORPORATION
- 3 -
MITSUBISHI CHEMICAL

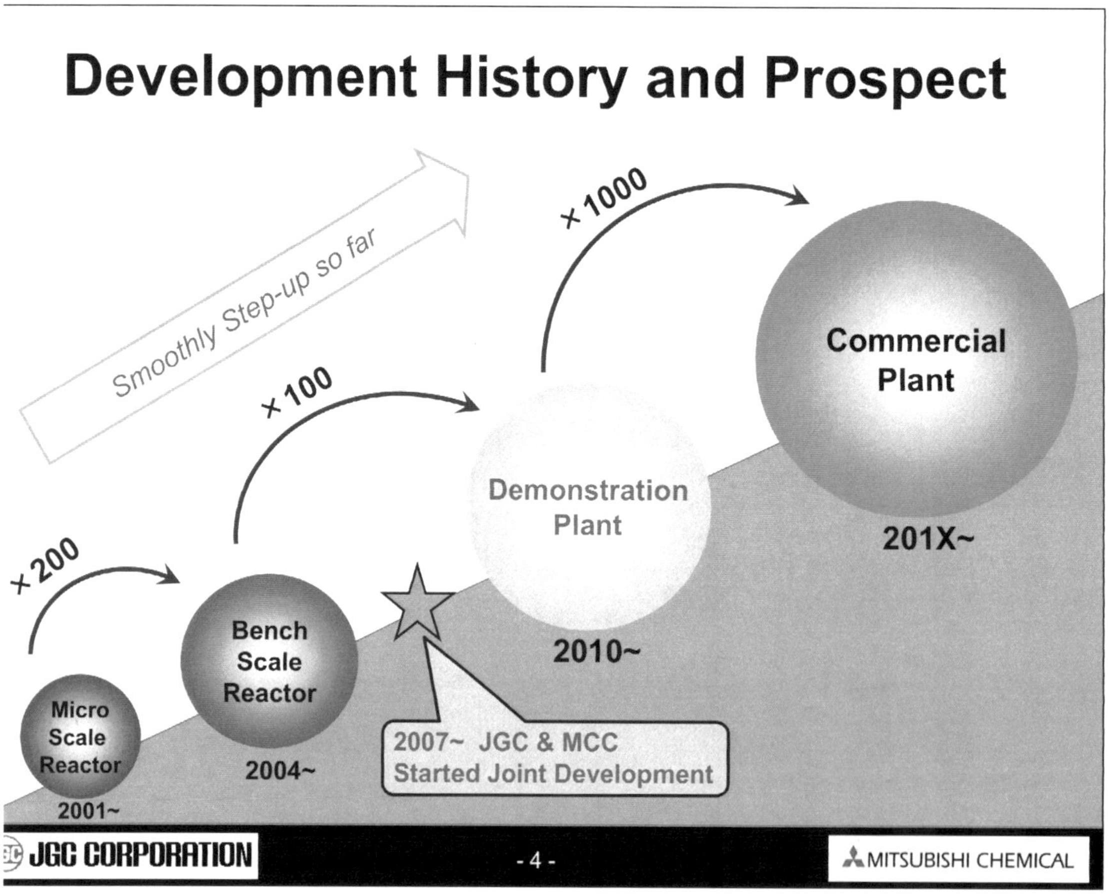
Development History and Prospect
Smoothly Step-up so far
×200
×100
×1000
Micro Scale Reactor
2001~
Bench Scale Reactor
2004~
Demonstration Plant
2010~
Commercial Plant
201X~
2007~ JGC & MCC
Started Joint Development
JGC CORPORATION
- 4 -
MITSUBISHI CHEMICAL

Demonstration Plant Overview

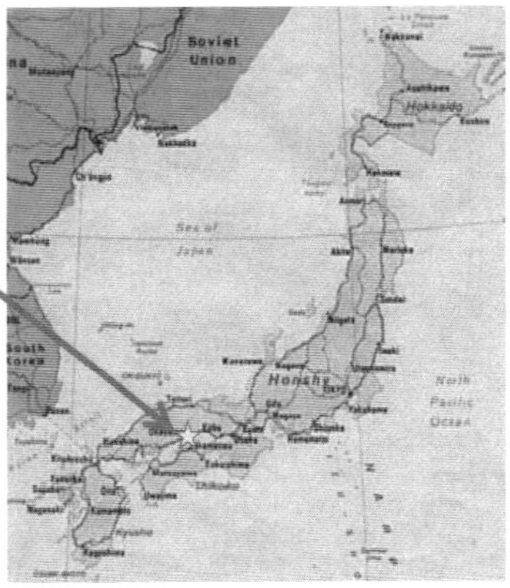

- **Located in Mitsubishi Chemical Mizushima Plant**
- **Successfully Start-up in August 2010**
- **Continuous Operation for Two Years**

Process Flow (Standalone Case)

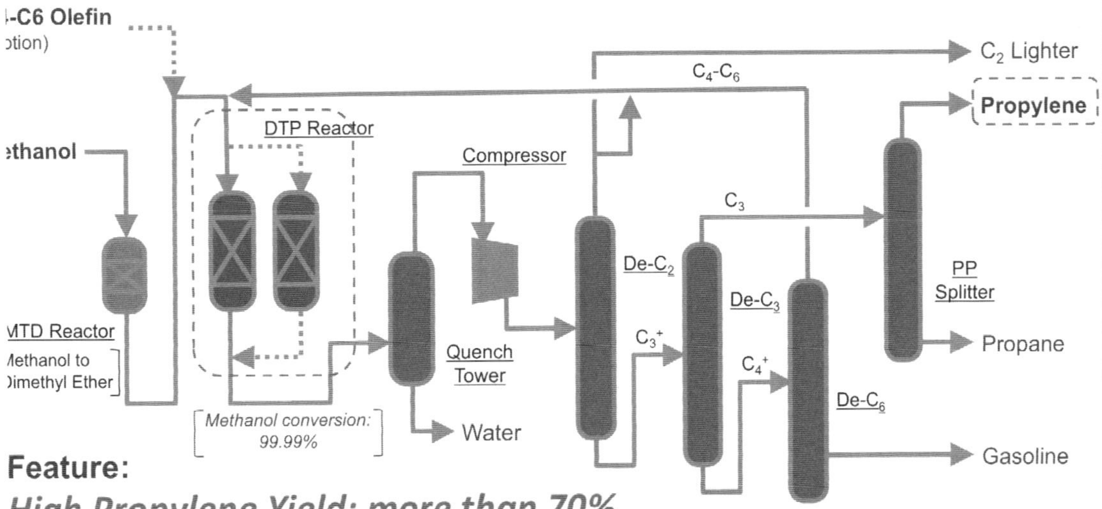

Feature:

High Propylene Yield: more than 70%

Applied Existing Proven Technologies

Process Flow (Integration Case)

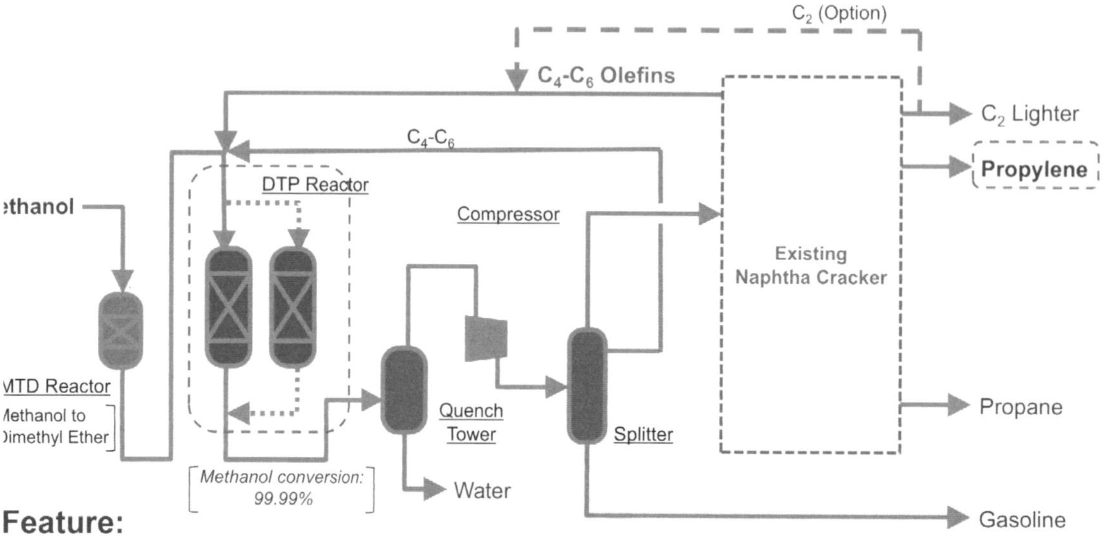

Feature:

High Propylene Yield: more than 70%

Applied Existing Proven Technologies

Low Investment Cost in case of Integration with Naphtha Cracker

Application

1) Naphtha Cracker Integration Case (Propylene Capacity: 20KTA – 200KTA)

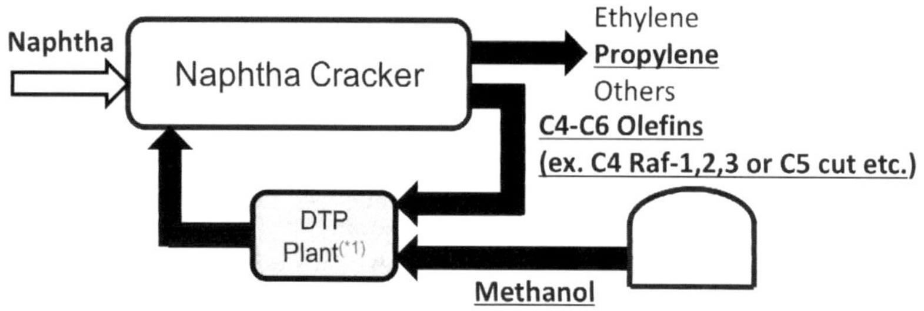

(*1) Utilizing facilities in Naphtha Cracker.

2) Standalone Case (Propylene Capacity: 250KTA – 500KTA)

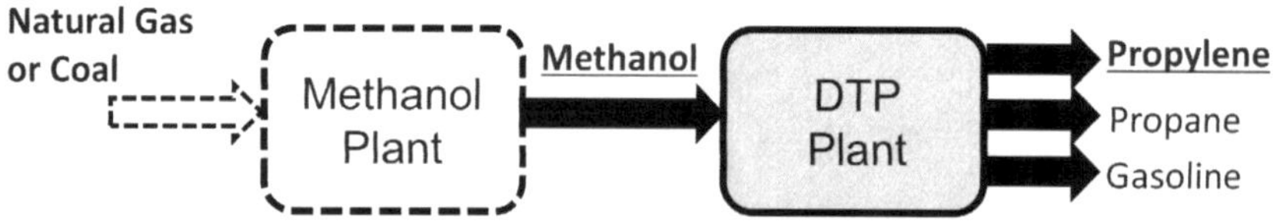

- Two Cases available depending on Client requirements !!

DTP is…

- *High Yield Propylene Production Process*
- *Flexible Feed Methanol to Olefin Process*
- *Process that increase P/E Ratio for Cracker*

Thank you very much !!

Contact:
Mitsuo Morita
Market Development Dept.
Global Marketing Division
2-3-1, Minato Mirai, Nishi-ku,
Yokohama 220-6001, Japan
Phone: 81-45-682-8371
FAX: 81-45-682-8183
E-Mail: morita.mitsuo@jgc.co.jp

352533 Self-Cleaning "Bernoulli" Type Filters Used in Onshore and Offshore Applications

Abstract:

Georg Schuenemann GmbH (SAB-BREMEN) has developed a new Filter Type F-480, dubbed the "Compact Giant" for applications in onshore water intake and cooling water systems, sea water desalination plants, and in offshore projects, such as FPSO's, Drillships, and Production Platforms.

This paper will describe the essential design features of the filters, based on the "Bernoulli principle" which result in a fully automated self-cleaning filter system, and choice of materials in a sea water service with high corrosion rates.

A case study will be presented in which stringent project requirements such as substantial flow rates, adaptibility to space limitations and pipe connections had to be observed. Reliability of the system, extended run times without the need for frequent replacement of filter elements, as well as ease of monitoring and operator involvement have been achieved.

352653 Effects of Support on Sulfur Tolerance and Regeneration of Pt Catalysts Measured By Ethylene Hydrogenation and EXAFS

Abstract:

We investigated the effect of support on sulfur tolerance and regenerability under reducing environments by measuring the rate of ethylene hydrogenation, combined with hydrogen chemisorption, and extended x-ray absorption fine structure (EXAFS). Alumina and titania (P25) supports were used to prepare Pt catalysts at 1 wt.%. The catalysts were tested after treatment in H_2S/H_2 at 250 °C followed by regeneration treatments in H_2 at 250 °C, 350 °C and 450 °C. The sulfided catalysts exhibited a 20-27 times decrease in the rate of ethylene hydrogenation and a 4-6 times drop in the Pt surface area. Regenerations up to 450 °C were unable to remove all the sulfur, as shown by the presence of Pt-S bonds by EXAFS. However, a partial recovery of the hydrogenation rate per mole of Pt was observed on sulfided Pt/Al_2O_3 after reduction at 450 °C, while the induction of strong metal support interactions (SMSI) at reduction temperature above 350 °C was observed on Pt/P25, regardless of the presence of sulfur. For Pt/P25, the reversal of the SMSI state together with sulfur removal by mild oxidation suggests that sequential reduction/oxidation treatments may be more effective in restoring the S-free state of TiO_2-supported catalysts.

352799 Improved Operational Efficiency and Reliability through Insulation Materials Selection

Abstract:

This paper will explore the implications of insulation materials selection on the energy efficiency and reliability for Ethylene production units. Considerations will include insulation properties, structural support, installation methods, installation and maintenance costs.

BOILER MACT COMPLIANCE: MERCURY IN FUEL GAS

PATRICK LAINE, WES YONGE, & VIC VICKERY
PEI, HOUSTON, TX 77043

Patrick Laine, Ph.D.
Chief Scientist
Houston, TX
850-866-6686
plaine@pei-tx.com

Wes Yonge, PE
Chief Engineer
Houston, TX
713-806-8396
wyonge@pei-tx.com

James 'Vic' Vickery
Chief Technology Officer
Houston, TX
281-536-0899
vvickery@pei-tx.com

1 Introduction & Objectives

For most ethylene producers the primary objective of this sampling and analysis effort will be to verify and "establish a mercury fuel specification through which gas-fired boilers that use a fuel other than natural gas or refinery gas may be considered Gas 1 unit" as identified in 40 CFR Part 63.7521.(f).

Measurement and Monitoring Requirements: Decision Rules

❑ If Hg concentration ≤ 50% of Gas 1 specification* then no further testing required

❑ If Hg concentration > 50% and ≤ 75% of the specification* then semi-annual sampling required

❑ If Hg concentration > 75% of the specification* then monthly sampling required

*Specification is **40 μg/Sm³**

The goal will be to execute a sampling and analysis campaign that will produce compliance quality data. The key to producing and validating compliance quality data is a well-designed and executed QA/QC protocol. At a minimum, a performance based QA/QC protocol is recommended that includes the following components:

- Sorbent Trap Breakthrough Bed:
  - ❑ Sorbent traps consisting of a minimum of two sorbent beds in series so that the trailing bed can be analyzed to verify that no breakthrough occurred from the leading bed.
- Duplicate Samples:
  - ❑ The sampling system should be configured with two trains to accommodate the collection of simultaneous duplicate sorbent trap samples, in parallel, providing a means of verifying precision in the sample collection process.
- NIST (National Institute of Standards and Technology) Traceable Spike:
  - ❑ Sorbent traps with a third bed (downstream) that is spiked with a NIST traceable spiked mass of mercury serves as a matrix spike and provides a means of verifying that components in the matrix (the fuel gas) are not affecting the trapping efficiency of the sorbent trap.

2 Methods

Sorbent Trap Mercury Sampling Methods Comparison

Method	Description	Method Type: S = Sampling A = Analysis	Evaluation	Sample Collection Phase QA/QC
ISO 6978	Hg in Natural Gas by Thermal Desorption (CVAFS)	S & A	Method calls for low sample volumes can lead to non representative results.	No
EPA Method 30B	Determination of Gaseous Hg Emissions from Coal Fired Combustion Sources using Carbon Sorbent Traps	S	Requires modification (Modified M30B)	Yes
ASTM D6350	Hg S&A in Natural Gas (AFS)	S & A	Method calls for short term samples (i.e., low sample volume) which leads to poor accuracy.	No
ASTM D5954	Hg S&A in Natural Gas (AAS)	S & A	AAS is less sensitive than AFS.	No
EPA Method 1631	Measurement of Hg in Water	A	No	Yes

3 EPA Method 30B QA/QC Performance Criteria

QA/QC Test or Specification	Acceptance Criteria	Frequency
Analyzer calibration	Correlation Coefficient: 0.999	Daily
Transfer efficiency of analyzer	%Transfer Efficiency: 90%–110%	Once per sorbent trap
Initial calibration blank	< 0.5 ng	Once before each calibration
Sorbent Trap Section B breakthrough	≤ 10% of Section A Hg mass, for Hg concentrations > 1μg/dscm ≤ 20% of Section A Hg mass, for Hg concentrations ≤ 1μg/dscm	At each sample point
Spike Recovery Test	Average recovery between 75% and 125 % for Hg_0	Once per sample point
Paired Sorbent Traps Agreement	≤ 10% RD for Hg concentration > 1μg/dscm ≤ 20% RD or 0.2 μg/dscm absolute difference for Hg concentrations ≤ 1μg/dscm	At each sample point

4 Equipment

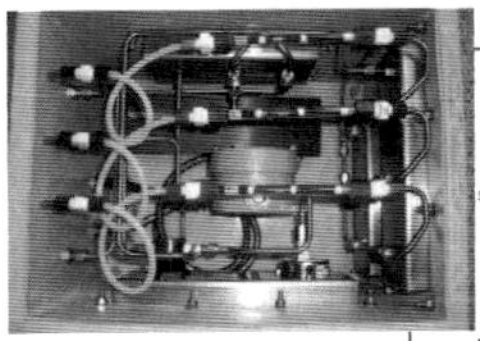

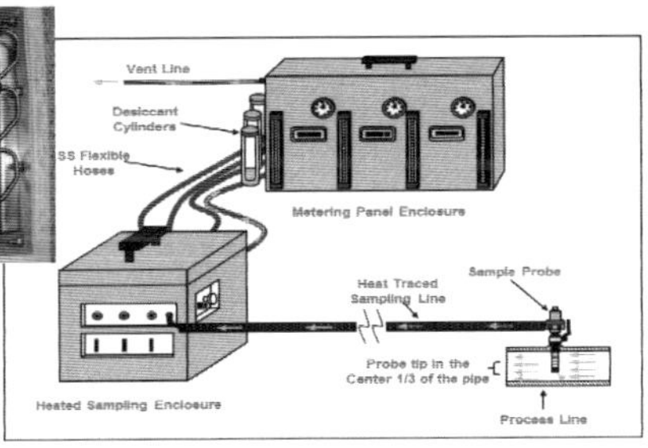

	Primary Bed	Breakthrough Bed	Spiked Bed
Flow			
QA/QC Acceptance Criteria:	99% of Total Collected Hg	≤ 10% of Primary Mass	Recovery between 75% and 125%

Gold Trap:

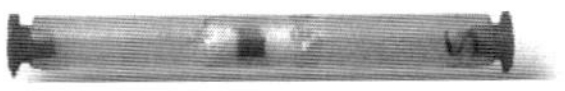

5 Conclusions/PEI Recomendations

❑ There are five methods accepted by the Boiler MACT rule. PEI has extensive experience and expertise employing **all five** of these methods.

❑ Operators can achieve compliance without submitting a Fuel Analysis Plan by using ASTM D6350, however, we have noticed variable results with this method.

❑ The **PEI recommendation** for obtaining the most accurate compliance quality data is to use Modified EPA Method 30B in conjunction with EPA 1631 where traps should be analyzed in a certified ISO 17025 laboratory. The recommended pathway includes a rigorous, performance based QA/QC protocol that validates accuracy and precision of all sample collection procedures.

351376 Young Professional Simulation Tutorial

Abstract:

18C00 Young Professional Simulation Tutorial

Authors: Naomi Hua and Mike Donahue, Invensys

Process simulation is a complex skill to master. Even with years of experience, a typical user is constantly facing new problems that need to be addressed. There are many issues that can lead to unexpected and unnecessary errors with the use of process simulation tools. Regardless of what simulation program you are using, these issues are common from one program to another. In this presentation, targeted for the young professional, we will discuss the "Most Common Pitfalls in Process Simulation". For each topic, we will discuss various techniques and methods these troublesome problems and prevent their occurrence.

Topics to be discussed include:

1. Erroneous data entry

2. Simulation defaults

3. Bad plant data

4. Thermodynamic method selection

5. Simulation tolerances and convergence
6. Recycle identification and handling
7. Model complexity
8. Difficult or impossible specifications
9. Assay pseudo-component development
10. Assay property blending

1. Erroneous data entry is the most common mistake made when building a process simulation model. This presentation will cover methods that can help ensure proper data entry in your process simulation model, such as checking the data you entered, particularly the units of measure, using "alternate forms" of data input such as copy/paste and data import, and reviewing all warning and error messages.

2. Defaults are methods, values, and conditions that are automatically selected for you by the program when you begin a new simulation model. These defaults will affect the results calculated by your model and every simulation program has them. Most defaults within simulation programs fall under categories such as convergence criteria, thermodynamic methods, tolerance and units of measure. The problem with defaults is unintended results from not fully understanding the effects a particular default has on the model. Take the time to understand the defaults of the program and how these defaults will affect the results. If you do not agree with the defaults selected by the program, most simulation programs will allow the user to change the default.

3. Plant data changes over time. The feed information and process conditions of the unit operations may be different. Not only does plant data change over time, but measurement information is often very inaccurate. This could be due to faulty or poorly calibrated measurement equipment, leaks in lines, sensor failures, and other issues. Because of these changes and inaccuracies over time, you always have to validate the information that is entered in your model to ensure that you are obtaining the proper results. To do so, you can utilize programs that can help with data validation and reconciliation like SimSci's ROMeo online optimization program. With it, you can connect your model to your data historian to constantly import current data measurements into your model and then reconcile these measurements to determine when problems arise within your plant and where those problems are occurring. You will also need to use your engineering judgment to determine if the data is valid. More details will be provided during the presentation.

4. The proper selection of thermodynamic methods for your simulation is the most important decision that you make when modeling a process. The method you select could have a drastic effect upon your results. Often, any number of thermodynamic methods is suitable for a given application. It is up to the user to determine the best method to use based upon knowledge, personal experience, and best judgment. Some important factors that need to be considered are the components being used and the operating conditions.

Oftentimes, the selected thermodynamic method still needs to have additional inputs such as binary interaction parameters to simulate the process accurately.

5. Many of the calculations in process simulation are iterative. When these calculations iterate, the program compares the results calculated from one iteration to the results calculated in the next iteration in order to determine

whether it is converged to a solution or needs to continue calculating. This method of comparing results is where tolerances come into play and simulation programs use tolerances to determine when a solution is reached. One of the major mistakes when modeling a process is not understanding or using the correct tolerance values. A looser tolerance may make the file easier and quicker to converge, but it also reduces the accuracy of the model. A tighter tolerance may be more accurate, but might make it impossible for the model to solve or increase the convergence time exponentially. A common practice is that when you have reached a converged solution with your model, tighten the tolerances within the model and solve it again. If you notice that your answers have changed significantly, you need to continue tightening the tolerance values until your results change little from one iteration to another. You will then know that you have reached an accurate solution for your model.

6. Recycles in process simulation occur when you return to a previously calculated unit operation to recalculate the unit due to a change made elsewhere in the model. They are very common and can often be difficult to converge. Some suggestions on troubleshooting recycle convergence include initially solving your model without the recycles to see how it behaves, providing initial estimates for recycle streams, using tighter tolerances on unit operations within recycle loops than the total recycle tolerances themselves, and ensuring that all components have a way in and out of a recycle by performing mass balances and using purge or makeup streams where necessary.

7. The general rule of thumb when modeling is to start simple and add complexity as you go. Each time you add a few streams or unit operations converge and save the model before moving on. Make sure you justify any additions that you make to the model. You can most likely model your system without including every unit from your PFD in the simulation. When modeling, think about the goal of your simulation and not about literally translating every real-world piece of equipment into the simulation model. Make the judgment call on what is most important and what the purpose is of every unit you add to your simulation. For example, if your real-life process shows a heat exchanger, a pump and a splitter, but all that is necessary is that the temperature, pressure, and separation are achieved properly, consider using a flash drum to accomplish all three.

8. Specifications within process simulation are everywhere and every stream and unit operation you place in your model contains specifications. In its simplest form, a specification is a constraint that you place on the model that the model needs to meet in order to successfully converge. For streams, these are usually temperatures, pressures, flowrates and compositions. For unit operations, they range from process conditions, like temperatures and pressures, to product conditions, like an overhead composition in a column. There are some pitfalls, however, that you can run into with specifications such as azeotropes in columns, high purity specifications, and the possibility of multiple solutions.

9. Petroleum assays are sets of distillation data that are used to characterize all of the components that reside in a given crude sample. When you enter distillation data into a process simulator, the program converts it into components that it can use when modeling your process. The components that are generated are called pseudo-components, and each pseudo-component represents all of the unknown components that would be within a given boiling point range in the crude. Some of the things you will want to ensure when properly developing your assay components are did you use the correct distillation curve basis, did you enter the data correctly, did you use enough cuts and are you using the correct characterization methods .

10. When you have more than 1 assay in a simulation model, you need to be concerned with assay blending and how it will affect the results of your file. Assay blending is the mixing of 2 or more groups of components (i.e. ¨C sets of cuts/pseudo-components) in order to calculate averaged component properties for cuts that are in the same boiling point range as one another. Once these new averaged component properties are determined, the program will use these averages from this point on in the simulation model any time it needs to access component

properties for a given pseudo-component. All component properties are blended, including normal boiling points, specific gravity, molecular weight and refinery inspection properties, and usually the blended component properties are calculated using a weighted average

These topics are based on experience and feedback from end-users to technical support representatives for process simulators. They often represent the most commonly asked questions or areas that most users find to be problematic. While the specific examples highlighted in each topic come from SimSci software, the intent of this tutorial presentation is to be broad-based and is particularly applicable to young professionals seeking to develop their experience with simulation.

Polymeric Sulfonic Acid Composites for Fuel Cell Applications: Effects of Casting Solvent and Sonication Time on Membrane Performance

Jimoh K. Adewole[1], *Abdullah S. Sultan*[1, *], *Amir Al-Ahmed*[2], *and S. M. Javid Zaidi*[3]

[1]*Center for Petroleum & Minerals, King Fahd University of Petroleum & Minerals, Dhahran 31261, Saudi Arabia*

[2]*Center of Research Excellence in Renewable Energy, King Fahd University of Petroleum & Minerals, Dhahran 31261, Saudi Arabia*

[3]*School of Chemical Engineering, The University of Queensland, Australia*

Abstract

Polymer blends comprising sulfonated poly (ether ether ketone) (SPEEK) and 3wt% (2-acrylamido-2-methyl-1-propanesulfonic acid) (PAMPS) were prepared by solution casting method. Two solvents (N,N-Dimethylacetamide (DMAc) and N-Methyl-2-pyrrolidone (NMP)) were uesd separately. Residual solvents in the membrane samples after drying were quantified using thermo gravimetric analysis. Effects of periods of sonication (0 to 90 minutes) and types of solvent on the membrane performance were evaluated interms of morphology and water uptake. An attempt was made to correlate membrane performance with the time of sonication in order to select the most suitable solvent and membrane casting conditions for subsequent membrane preparations. For both solvents, the water uptake was observed to first decrease at 10 minutes of sonication and then increased as the sonication time was increased. The minimum water uptake recorded for DMAc samples was 19.23% at 20 minutes sonication time while that of NMP sample was 24.24% at 10 minutes. However membrane morphology was found to deteriorate with increase in sonication time. No correlation was obtained between the membrane properties and the sonication time. In general, membrane performance was better for membranes samples using DMAc than NMP solvent. Therefore, DMAc was selected as the casting solvent for membrane preparation on which further studies will be carried out.

Keywords: morphology, water uptake, residual solvent, thermogravimetric analysis

Introduction

Water soluble polymers have received academic and industrial interests due to their potential applications in fuel cells and other electrochemical devices (Ye et al. 2012). One of the water soluble polymers that have attracted publication interest is poly (2-acrylamido-2-methyl-1-propanesulfonic acid) (PAMPS) (Florjańczyk et al. 2003). Generally, some of the major performance criteria for fuel cell materials include proton conductivity, water and methanol uptake, good mechanical properties, chemical stability, thermal stability and methanol permeability (Gil et al. 2004; Swier et al. 2005; Ye et al. 2012). PAMPS was reported to posses higher proton conductivity than partially hydrated Nafion. In addition, it has been found to be more tolerant to water fluctuation in high temperature operations. Some of the major drawbacks in the use of PAMPS are its poor mechanical properties, thermal stability, water and methanol uptake and solubility in many solvents (Florjańczyk et al. 2003). One of the approaches that have been proposed to solve these problems is development of polymer composites.

This research paper investigates the synergistic effect of combining water soluble PAMPS with sulfonated poly (ether ether ketone) (SPEEK) which is non - water soluble polymer to prepare composite membranes for application in fuel cells. Specifically, preliminary results on the performnance of the composite were presented. Performance were evaluated based on the composite morphology and water uptake. Effects of sonication time and types of casting solvent on membrane properties were also evaluated.

Material and Method

Material

Sulfonated poly (ether ether ketone) (SPEEK) purcahsed from FuMA-Tech, Germany, and poly (2-acrylamido-2-methyl-1-propanesulfonic acid) (PAMPS) purchased from Aldrich Chemical Co., are the two major polymers used in this research. The SPEEK has a DS of 0.533 and solubility of 99% in DMAc. SPEEK is derived by sulfonation of Poly (ether ether ketone) (PEEK). PEEK is a commercially successful polymer due to its excellent thermal stability and solvent resistance (Mohr et al. 1991). For sometimes, further application of PEEK have been stalled due this solubility issues. By sulfonation, PEEK can be made soluble in most commercial solvents. Importantly, the solubility of SPEEK also depends on its degree of sulfonation (DS). Since, this work involves the use of two polymers, a solvent which will disolve both the PAMPS and SPEEK must be selected. Literature review showed that PAMPS dissolves in N,N-Dimethylformamide (DMF), N,N-Dimethylacetamide (DMAc), propylene carbonate (PC), and water (Florjańczyk et al. 2003). Initial solubility studies that was carried out in our laboratory was done for N,N-Dimethylacetamide (DMAc) and N-Methyl-2-pyrrolidone (NMP). The choice of NMP was due its unique solvent properties and its structtural distiction from DMF and DMAc (Figure 1). Moreover, no literature was found where NMP was used as casting solvent for PAMPS. Thus, DMAc and NMP were used as solvents in this work. All materials were used as received from the manufacturers.

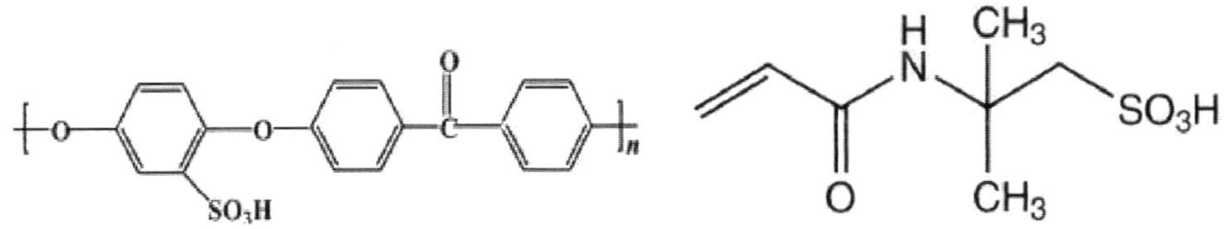

Figure 1 Structure of SPEEK and PAMPS

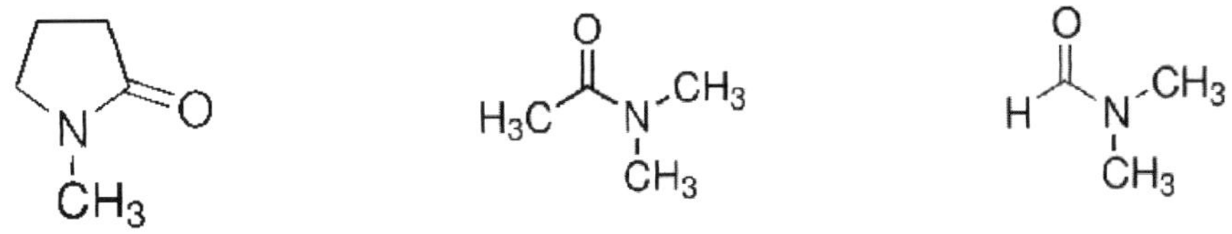

N-Methyl-2-pyrrolidone N,N-Dimethylacetamide N,N-Dimethylformamide

Figure 2 Structure of Solvents

Experimental Procedure

Membrane Preparation

SPEEK was first dissolved in solvents (DMAc and NMP) and stirred for 2hours at 70°C. PAMPS was then added and the blend was stirred for 24hours. The solution was sonicated for various periods of time ranging from 10 to 90minutes using an Ultrasonic bath. In untralsonic bath, the final temperature achieved for a given time of sonication depends on the tempearutre of the laboratory (summer or winter). It is therefoe genrally advisable that maximum temperature of the bath is determined and maintains (as the so - called equilibrium tempearture), when operating continuously under ambient conditions (Santos et al. 2009). During the sonication, equilibrium temperature was maitained by simply filling the bath with water that was preheated to sample temperature.

Membrane Characterization

Water uptake and morphological tests were performed. Water uptake was carried out by first vacuum drying all membrane samples at 100°C until constant weights were obtained. Dried samples were then immersed in deionized water for 24h at room temperature (25°C). Samples were taken out of the water after 24h, wipped to remove surface water and then weighed immediately. Water uptake was calculated using Eq. 1.

$$water - uptake = \frac{W_{wet} - W_{dry}}{W_{dry}} \qquad \text{Eq. 1}$$

Surface morphology of the membrane was done using scanning electron microscopy (SEM). Thermogravimetric analysis (TGA) was done from room temperature to 350°C using TA Q600 analyzer at heating rate of 20°C/min under nitrogen atmosphere with flow rate of 20mL/min using ~ 8 -10mg membranes samples.

Results and Discussion

Membrane Morphology by Scanning Electron Microspcopy (SEM)

Results of surface morphological analysis performed on membrane were shown in Figure 3 and 4. The overall pictures of this figure depicts a dense membrane with some surface defects observed for most of the sonicated samples. For example, Figure 3A showed some kind of layered arrangement with no visible pores at the surface. Also, the PAMPS seems to be well distributed throughout the SPEEK. The micrograph in Figure 3B can be observed to be rough and possesses some uneven ramdom distribution of PAMPS agglomerates indicating that the distribution of PAMPS have been disturbed by the 10minutes sonication. An increase in sonication time to 20 minutes was observed to improve the distribution of PAMPS. This was however accompanied by some cracks believed to be due to the slight increase in the viscousity of the blend solution. The vicousity increase was brought about by the excape of some fraction of the solvent during sonication. As the sonication time was increased, PAMPS was observed to become more evenly distributed and the cracks become more visible and larger in size (Figure 3D to F). SEM images of samples at sonication time of 10 and 30 minutes were reproduced in higher magnification of 10μm (Figure 4) in order to further revealed the effect of sonication on the morphology of the somposites. Similar morphological behaviour has been reported in literature, though for inorganic/organic clusters of nanoclay/epoxy composites membranes (Lam et al. 2005).

Water Uptake

Water uptake have been reported to be affected by the degree of sulfonation, membrane pretreatment, hydration, ambient relative humidity and water temperature (Xing et al. 2004; Wu et al. 2006). Figure 5 showed the effect of sonication time and solvent type on water uptake that was measured at room temperature (25°C). For both solvents, water uptake is same at zero sonication time. There was however a further decrease in the water uptake for NMP at 20 minutes sonication time before it beging to increase. Thus, the best sonication time for both solvent lied between 10 and 20 minutes with respect to water uptake. Interestingly, this result correspond to that reported by Lam (2005) on the effect of ultrasonication on microhardness of composites.

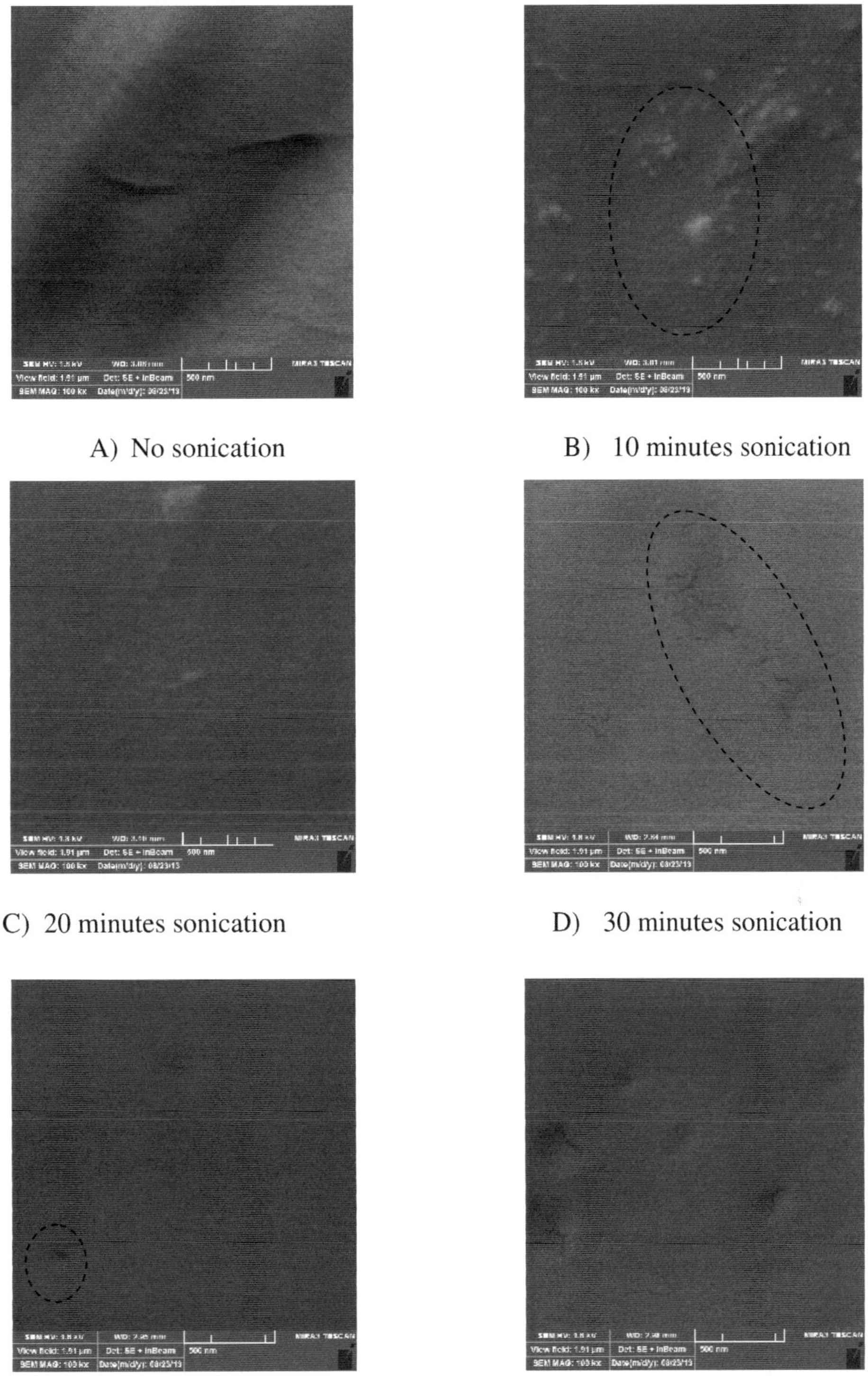

A) No sonication
B) 10 minutes sonication
C) 20 minutes sonication
D) 30 minutes sonication
E) 60 minutes sonication
F) 90 minutes sonication

Figure 3 SEM micrograph of SPEEK-PAMS composite with various times of sonication with magnification of 500nm

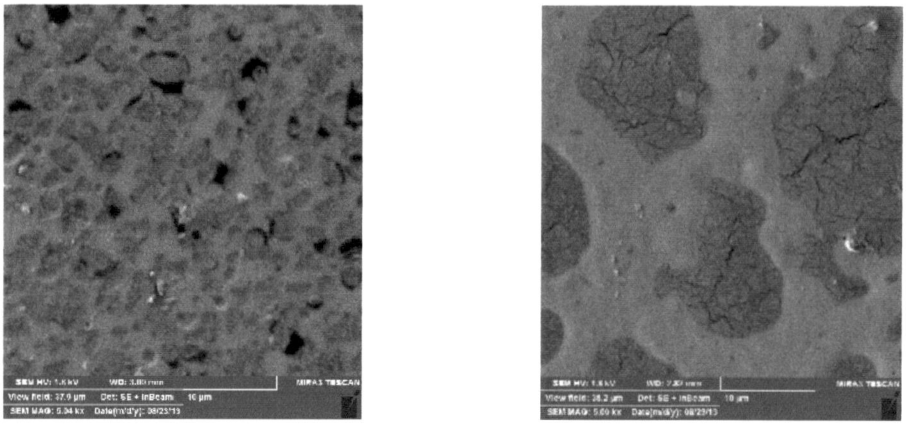

A) 10 minutes sonication B) 30 minutes sonication

Figure 4 SEM micrograph of SPEEK-PAMS composite with various times of sonication with magnification of 10μm

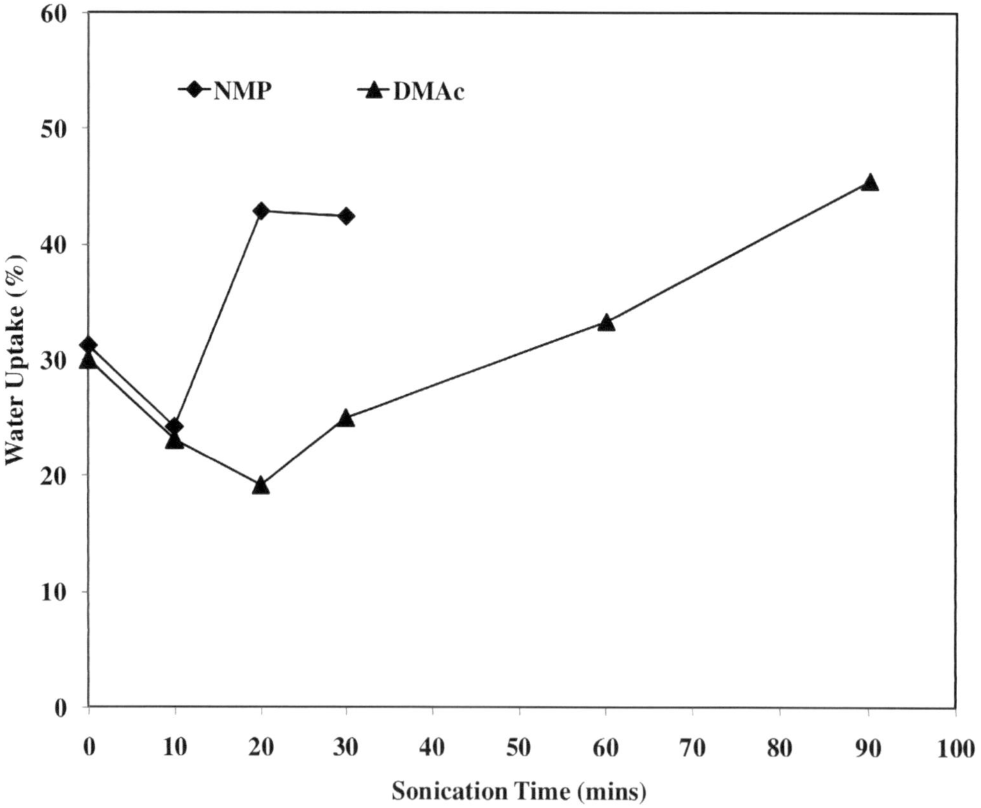

Figure 5 Water uptake of SPEEK-PAMS (3wt%) composite with variuos sonication times

Thermogravimetric analysis (TGA)

The difference in membrane performance resulting from the use different casting solvents have been attributed to the presence of residual solvents in the membrane samples after evaporation (Joly et al.

1999). TGA was thus used to quantify the residual solvent and evaluate its effect on the membrane performance as has been previously demonstrated in lieterature (Recio et al. 2007; Iqbal et al. 2008). Figure 6 shows the TGA curve for samples prepared from NMP and DMAc with different post vacuum heat treatment conditions. From this figure, weight loss can be attributed to evaporation of variety of substances, including solvents, from the samples (Rich et al. 1997). Table 2 contains the solvent and its boiling points, and weigth loss at the boiling point range that were extracted from TGA results. For sample heat treated for 24hrs, the overall percentage weight loss up to 204°C was about 8wt% of membrane sample for NMP while it is about 6.3wt% for DMAc. These will definitely include compnenents such as water, solvents and other unidentified substances. To quantify residual solvents, weight loss was estimated between 192 -214°C for SPEEK-PAMPS-NMP and between 154 and 176 for SPEEK-PAMPS-DMAc. The results obtained showed that the amount of the residual solvent in NMP is larger than that of DMAc by 1.25wt% (Table 1). To further investigate if this difference is responsible for higher value of water uptake for the NMP sample, membrane samples made from both solvents were further heat treated under vacuum for 72hrs. Residual solvent after this treatment was slightly higher in DMAc than NMP. Despite this, the value of water uptake of the NMP membrane is still higher than that of DMAc by approximately the same value as before the heat treatment (1.71wt%).

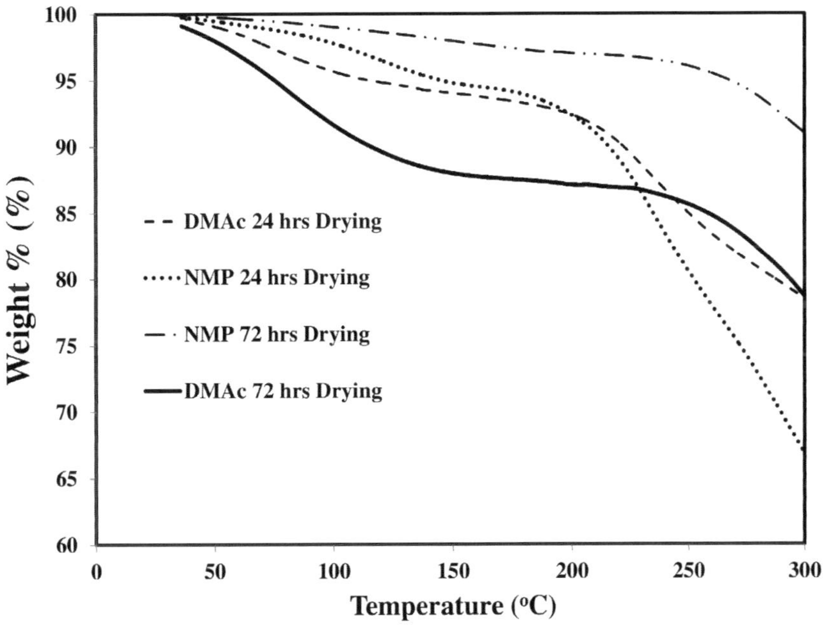

Figure 6 TGA of SPEEK-PAMPS-NMP and SPEEK-PAMPS-DMAc

Table 1 Residual solvent in SPEEK-PAMPS composites membrane samples

Membrane Sample	Solvent	B. Pt. (°C)*	Residual Solvent (wt%)	Water Uptake
DMAc 24 hrs Drying	DMAc	164.5 – 166	0.57	30.00
NMP 24 hrs Drying	NMP	202 – 204	2.85	31.25
DMAc 72 hrs Drying	DMAc	164.5 – 166	0.367	21.21
NMP 72 hrs Drying	NMP	202 – 204	0.217	22.92

*Boiling Points of Solvents

On a general note, results from Figire 1 – 5 showed that the best perfomance can be obtained for variety of sonication time depending the property of preference. Also, performance of DMAc solvent was observed to be better than that of NMP on the basis of water uptake. For this reason, DMAc was selected as the casting solvent for membrane preparation on which further studies will be carried out. Water uptake of DMAc membrane, prepared without sonication, was comapred with those reported in previuos literature. Table 2 illustrates the comparison between SPEEK-PAMPS and other relevant membranes previuosly reported in literature. In addition, this table serve to provide insight into the possible potentials of the composites. From this table, it was observed that the water uptake for SPEEK-PAMPS is within the range of values of other membranes on the table. Its water uptake is less than that of Nafion 112 which is a commercial membrane. The differences observed between the water uptake values of SPEEK – PAMPS and other membranes from the table can be explained based on three important factors - SO_3H group, the degree of sulfonation and experimental temperature (Gil et al. 2004; Shan et al. 2006; Wu et al. 2006). Water uptake of SPEEK – PAMPS is higher due to the presence of SO_3H group in the PAMPS molecules (Figure 2). SO_3H group have been reported to enhance membrane acidity and hydrophilic properties (Shan et al. 2006). Infact, the higher the SO_3H group the larger the water uptake. The degree of sulfonation has been reported to affect the water uptake of membranes. The higher the degree of sulfonation, the larger the value of water uptake (Gil et al. 2004). Thus, water uptake of SPEEK – PAMPS is lower than SPEEK -3 due their different degree of sulfonation, 53.3 and 72% respectively and the experimental temperature 24 and 60°C, respectively. The experimental temperature of SPEEK – 2 is close to that of the composite. However, the results cannot be compared in this case since the DS of SPEEK – 2 was not known. Above explanation is also true for the values of water uptake for SPEEK – 1.

Table 2 Comparison of water uptake values obtained with those from Literature (Xing et al. 2005)

Membrane	Water Uptake (wt%)	Experiment Temperature (^{o}C)	**D S (%)	Reference
Nafion 112	22.85	N/A	-	(Xing et al. 2005)
SPEEK – 1*	16.00	30	45	(Ren et al. 2005)
SPEEK – 2*	44.90	25	N/A	(Zhong et al. 2008)
SPEEK – 3*	85.00	60	72	(Wu et al. 2006)
SPEEK– 4*	15.00	N/A	N/A	(Handayani and Dewi 2011)
SPEEK/PAMPS (3wt%)	21.21			This work

*The numbering have been included for explanation purposes.
**DS means Degree of Sulfonation

Conclusion and Recommendations

Blend of SPEEK and PAMPS was prepared by solvent casting method and its performance was evaluated with respect to its morphology and water uptake. Effect of sonication time (10 – 90minutes) and solvent types (DMAc and NMP) were also investigated. Performance of membrane prepared using DMAc was observed to be better than that of NMP. There was no correlation between the sonication time and membrane properties, but it was observed that the sonication deteriorate the membrane morphology. Membrane prepared without sonication shows better surface morphology, whereas the best water uptake was observed between 10 and 20 minutes. Thus, other properties need to be investigated in order decide the best sonication time. Therefore, further investigation such as thermal stability, methanol permeability, ion exchange capacity and gas (O_2 and H_2) permeability of the membrane are recommended for a complete evaluation of the membrane performance for direct methanol fuel cell (DMFC) application. All these investigations are in progress in our laboratory.

Acknowledgement

Authors would like to thank the National Science, Technology and Innovation Program (NSTIP) of KACST for funding this research under project number 10-ENE1374-04. Also, authors extend their appreciation to Center of Petroleum & Minearls, Center of Excellence in Renewable Energy, Mr. Dan Jerry D.Cortes of the Nanotechnology Laboratory, Chemical Engineering Department, King Fahd University of Petroeleum & Minerals, Saudi Arabis for their supports.

REFERENCE

Florjańczyk, Z., et al. (2003). "Proton conducting electrolytes based on poly(2-acrylamido-2-methyl-1-propanesulfonic acid)." Electrochimica Acta **48**(14–16): 2201-2206.

Gil, M., et al. (2004). "Direct synthesis of sulfonated aromatic poly(ether ether ketone) proton exchange membranes for fuel cell applications." Journal of Membrane Science **234**(1–2): 75-81.

Handayani, S. and E. L. Dewi (2011). "Influence of Silica/Sulfonated Polyether-Ether Ketone as Polymer Electrolyte Membrane for Hydrogen Fueled Proton Exchange Membrane Fuel Cells." Internat. J. of Sci. and Eng. **2**(2): 26-30.

Iqbal, M., et al. (2008). "Solvent effect on morphology and CO2/CH4 separation performance of asymmetric polycarbonate membranes." J. Membr. Sci. **318**(1–2): 167-175.

Joly, C., et al. (1999). "Residual solvent effect on the permeation properties of fluorinated polyimide films." Separation and Purification Technology **16**(1): 47-54.

Lam, C.-k., et al. (2005). "Effect of ultrasound sonication in nanoclay clusters of nanoclay/epoxy composites." Materials Letters **59**(11): 1369-1372.

Mohr, J. M., et al. (1991). "Gas transport properties of a series of poly(ether ketone) polymers." Polymer **32, 13**: 2387-2394

Recio, R., et al. (2007). "Gas separation of 6FDA–6FpDA membranes: Effect of the solvent on polymer surfaces and permselectivity." J. Membr. Sci. **293**(1–2): 22-28.

Ren, S., et al. (2005). "Surface modification of sulfonated poly(ether ether ketone) membranes using Nafion solution for direct methanol fuel cells." Journal of Membrane Science **247**(1–2): 59-63.

Rich, D. C., et al. (1997). "Effect of cure conditions on Probimide 32 polyamide-imide." Journal of Applied Polymer Science **63**(9): 1113-1126.

Santos, H. M., et al. (2009). The Power of Ultrasound. Ultrasound in Chemistry: Analytical Applications. J.-L. Capelo-Martinez. Weinheim, Wiley-VCH.

Shan, J., et al. (2006). "Sulfonated polyether ether ketone (PEEK-WC)/phosphotungstic acid composite: Preparation and characterization of the fuel cell membranes." Pure Appl. Chem **78**(9): 1781 - 1791.

Swier, S., et al. (2005). "Sulfonated poly(ether ketone ketone) ionomers as proton exchange membranes." Polymer Engineering & Science **45**(8): 1081-1091.

Wu, H.-L., et al. (2006). "Preparation and characterization of poly(ether sulfone)/sulfonated poly(ether ether ketone) blend membranes." European Polymer Journal **42**(7): 1688-1695.

Xing, D. M., et al. (2005). "Characterization of Sulfonated Poly(Ether Ether Ketone)/Polytetrafluoroethylene Composite Membranes for Fuel Cell Applications." Fuel Cells **5**(3): 406-411.

Xing, P., et al. (2004). "Synthesis and characterization of sulfonated poly(ether ether ketone) for proton exchange membranes." Journal of Membrane Science **229**(1–2): 95-106.

Ye, Y.-S., et al. (2012). "Water Soluble Polymers as Proton Exchange Membranes for Fuel Cells." Polymers **4**(2): 913-963.

Zhong, S., et al. (2008). "Modification of sulfonated poly(ether ether ketone) proton exchange membrane for reducing methanol crossover." Journal of Power Sources **180**(1): 23-28.

Fixed Plant Facility Siting – A Mega Project Lifecycle Framework

M.F. Haider
mech_haider@yahoo.com
mhaider@suncor.com

Prepared for Presentation at
American Institute of Chemical Engineers
2014 Spring Meeting
10th Global Congress on Process Safety
New Orleans, LA
March 30 – April 2, 2014

Fixed Plant Facility Siting – A Mega Project Lifecycle Framework

M.F. Haider
mhaider@suncor.com
mech_haider@yahoo.com

Keywords: Facility Siting, Project Management, Mega Project, Consequence Analysis (CA), Quantitative Risk Assessment (QRA), Adaptive, Change driven, Agile methodology

Abstract

Facility Siting based on tools such as Consequence Analysis (CA) and Quantitative Risk Assessment (QRA) is often times successfully applied for Brownfield operating unit projects, projects with relatively smaller footprints and for projects lasting relatively short time-periods. Challenge remains to apply and realize the full potential and value of Facility Siting for Mega-Projects. This is mostly due to project complexity and length – this paper will introduce a Management Framework to successfully apply CA and QRA as part of Facility Siting for Mega-Projects. The paper will also outline key strategies to ensure successful (i.e. value adding) implementation of facility siting, suggest a list of activities, associated sequence and pitfalls to avoid. Project Management team (e.g. Project Sponsor, Project Engineers, Managers) of both owner/operator and contractor/vendor may find this paper more useful as it provides a broad overview (i.e. when to initiate, how to plan, execute and close such assignment to maximize value) including real life examples of facility siting process over the project lifecycle.

1. Introduction

The driving force for process safety has been primarily based on catastrophic events [1]. With an increasing number of incidents, the process industry and governments started to take measures such as Facility Siting initiatives with hope to minimize loss of life and property. In the 1970s and 1980s, some of the world's most tragic industrial accidents took place. The Flixborough explosion in 1974 was reinforced by that of the Seveso tragedy in 1976 [2]. The Macondo blowout in the Gulf of Mexico on April 20, 2010, killed 11 employees and led to an uncontrolled oil spill lasting 87 days. This blowout was the most significant offshore incident in the US to-date. Finally, the Fukushima Daiichi nuclear plant incident in Japan in March 2011 drew the attention of the global process and power industries and encouraged them to incorporate natural disaster risks in a hazard analysis study [3]. As part of inherently safer design, many companies now-a-days apply CA and QRA as part of facility siting for their new projects. Due to the scarcity of information at the beginning, the studies in many cases start later in the project lifecycle for them to be effective and value adding. This paper will focus more on how to manage such initiatives for new Mega-Projects over the technical intricacies of the CA/QRA processes. The intention of the paper is to enhance the Body of Knowledge publicly available for management team in as much non technical languaga as possible to make their greenfiled Mega-Project initatives a success i.e. safer for people, environment and property.

So, what is different for industrial Mega-Projects?

- Industrial Megaprojects, as the name implies, are very large. To provide a simple and simply applied definition, we are defining a megaproject as any project with a total capital cost of more than $1 billion (U.S. dollars) as measured on January 1, 2003 [4].
- Most often engineering and construction activities run in parallel and not in sequence
- Multiple sub-projects under the 'Mega-Project umbrella' run in parallel
- Complex – design is almost 'never' complete in early stages – goes through many iterations
- Progressive elaboration is evident
- Scarcity of critical workforce with mega-project experience
- Scarcity of information – requires project team to make assumptions for CA/QRA
- Timeline: 2-5 years

2. Key Strategies

2.1 Facility Siting Charter

It is not only important or good to do but critical that the stakeholders need to identify, understand and agree on the purpose and value to conduct facility siting for their projects. The Author's personal experience over the years has been in order to maximize value of Facility Siting the project team has to start 'early' to develop project Charter and start with the end in mind. As can be seen from the following generalized Figure 1 that as the company starts to appraise opportunities, it is

Figure 1: Stages and Milestones to apply CA/QRA in a Mega-Project

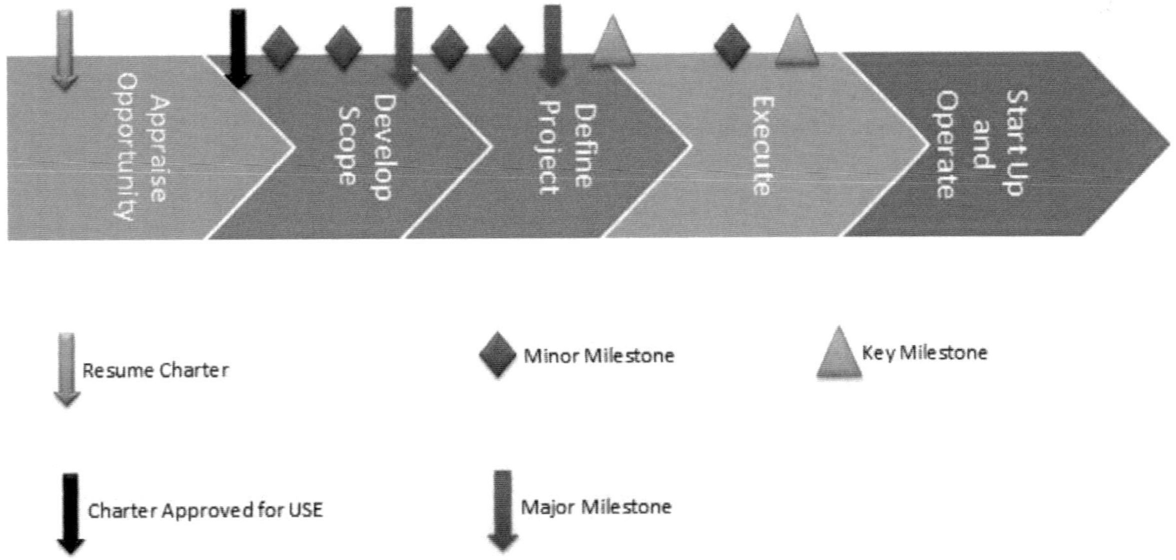

prudent to resume developing the Facility Siting Charter by bringing in the stakeholders together. Figure 1 shows a generalized project lifecycle for a typical Petro-Chemical Mega-Project. The

Figure also shows that the Key Deliverable is developed through a successive series of repeated cycles over the project life-cycle. The Milestones will be defined later in the paper. Part of starting with the end in mind is to clearly outlining ultimate goal/s, e.g. vision is to design a safer facility and utilize Mitigated QRA Report to formally document and communicate that facility is within the Risk Threshold or justification for accepting risk profile as is by Senior Management. The Vision in the Charter also provides opportunity to strike a balance between business goals and climbing the Risk Mountain – at each stage of increasing safety (decreasing risk). As part of striking a balance, the Project Engineer has to Identify options up-front if and/or when it becomes apparent that aggregate Residual Risk will not be within the Risk Threshold or acceptable so that team can put more effort in developing justification for either accepting or rejecting the risk profile by Management

Figure 2: Strike a balance between Business goal and Climbing the *expensive* Risk Mountain

2.2 *Management Strategy*

In order to successfully manage Facility Siting, CA and QRA for a Mega-Project, the stakeholders have to align on a management strategy. There are two primary management approaches for the purpose of managing such work:

1. Predictive or Plan driven
2. Adaptive or Change driven

The Author's past experience has been if the organization undertook very 'similar' projects in the near past (<5 years), critical experienced workforce is retained, Lessons Learned is readily available, and most importantly, the past project is still considered to be a success, it is advisable to use Plan driven approach. On the other hand, if the organization is introducing new technology, has limited past experience with similar project, involves relatively high number of stakeholders, capital cost > $1B, project life spans >2 years, it is recommended to utilize Change driven approach to maximize value. It is to be noted that it is not essential to meet all the suggested criteria mentioned above to pick one management strategy over the other.

2.2.1 Benefits of Adaptive or Change Driven Approach

In Adaptive/Change Driven Planning, project team intentionally repeats one or more project activities as the project team's understanding of the product increases ....Also known as ...agile methods .....intended to respond to high levels of change and ongoing stakeholder involvement [5]. The heart of Facility Siting involves making timely but informed decision. Some of the characteristics of this framework:

1. To add small incremental improvements that will deliver value to stakeholders
2. Milestones develop the product (e.g. Mitigated QRA Report) through a series of repeated cycles, while increments successively add to the functionality of the product
3. For each Milestone, a deliverable (read 'model') or set of deliverables will be completed. Future iterations may enhance those deliverables or create new ones. Each iteration incrementally builds the deliverables until the exit criteria for the phase are met, allowing the project team to incorporate feedback.

The biggest challenge with Mega-Project is scarcity of key information needed as inputs to develop scenarios and models. Some of the critical benefits, a project may gain by applying Change Driven approach:

1. The approach acts as Leading Indicator on Process Safety for Plant Layout - allows for multiple opportunities to make critical decisions on plant layout over the project life-cycle capitalizing on what is known at a given time. This lets the project team to use "what they know", to ensure "what they don't know" falls into the right place
2. Opportunity to make decisions over the project lifecycle not only works to save life but also helps with striking a balance between business needs and climbing the risk mountain
3. Aides design-development with progressive elaboration while producing final CA/QRA deliverable for the project

2.3 Execution Strategy

The main idea of Change Driven process starts with decomposing the known preliminary layout and process information into a set of requirements, work to be performed and their priority sequence, sometimes referred to as a *Product Backlog*.

It is advisable that the owner/operator forms a multi-use agreement (contract) with the chosen vendor. This way the plant areas under the Mega-Project umbrella can get their specific work done while the CA/QRA Project Team provides overall oversight and guidance.

At the beginning of an iteration, the team will work to determine how many of the highest priority items on the backlog list can be delivered within the next iteration. The time-length of each iteration should be less than 4 weeks. As shown in Figure 1, during "Develop Scope" and "Define Project", the length of each iteration could be even less than 2 weeks. This iterations (otherwise known as "Minor Milestone" as shown on Figure 1) will develop scenarios, assumptions and preliminary CA/QRA models. Assumptions are required due to lack of available quality information in the early stages of a project.

At the end of each iteration, the Minor Milestone/deliverable should be ready for review by the customers/stakeholders. This does not mean that the customer is required to accept delivery, just that the product should not include unfinished and/or incomplete features.

The sponsor and customer representatives should be continuously engaged with the other project stakeholders to provide feedback to Contractor on deliverables as they are created and to ensure that the *Product Backlog* reflects Sponsor/stakeholders' current needs. Below, you will find two examples using two different management strategies. For confidentiality purpose, the specific name/location is kept generic and few items were modified to suit.

2.3.1 Project X Facility Siting Using Change Driven Approach

A CA/QRA team was given the assignment to support design development during "Develop Scope" phase (Figure 1). The Sponsors already selected the site for the project. But the fixed plants were yet to be located on the site. The Vision for the initiative was for the QRA result to come under the organization Risk Threshold. Multiple EPC's were given to engineer different plant areas. The CA/QRA team selected a vendor to support them with defining the *Product Backlog* based on the preliminary Process Design Basis, intended product the facility is expected to sell, the feedstock necessary etc. The CA/QRA team had prioritized the Product Backlog based on hazard level of feedstock, preliminary PHA schedule and project schedule. This also involved obtaining feedback from different Process Engineering and Operations teams to collect necessary inputs to put together the assumptions and scenarios to be modeled within the 1st iteration.

Out of 25+ scenarios, only 7 perceived to be the most hazardous were modeled for CA that gave ample input (e.g. explosion, toxic and fire contours) for the CA/QRA team to work with the stakeholders to recommend location for specific process units on the given site. Note that, CA/QRA team integrated the potential consequence to understand local incident and global consequence events. At this time QRA was not performed. This was used as input for minor modification of preliminary plant layout. It is to be noted that the site plan was still preliminary after the first iteration but assisted different EPC accountable for their specific area to recognize where to locate the specific plants i.e. finalized key plans for the overall site plan.

With further design-development, more refined information became available from each EPC. This new information led to 130+ scenarios and was used for further CA/QRA iterations that created either new model and/or embellished the past ones. Each EPC was accountable for facility siting within their boundary while the CA/QRA team provided technical oversight and integration support for the overall project. The Sponsor/stakeholders were engaged on a regular basis to ensure that the Product Backlog reflects future modeling requirements to properly site facilities e.g. Fire Water Pump house, Tank Farms, site Medical Centre etc. Early engagement of CA/QRA team and higher management willingness to incorporate their feedback with the progressive elaboration of the project led to QRA result meeting the organization risk threshold and ultimately safer facility. Changes suggested based on CA/QRA work in earlier stage (i.e. Develop Scope, Define Project) were easier to incorporate into the design.

2.3.2 Project Y Facility Siting Using Plan Driven Approach

The sponsors for project Y selected the site for the project and decided to hold on to not perform CA/QRA work with the progressive elaboration of the project activities. This was mostly decided based on the fact that not all information was available to properly conduct the 'studies'. Fundamentally, the sponsors looked CA/QRA as deliverables requiring a more 'complete' plan. Once the Develop Scope (as referred in Figure 1) is almost complete, the sponsors decided to engage CA/QRA team and a vendor. By this time, the project had significantly high number of stakeholders. Many of the stakeholders were not very familiar with CA/QRA and the value it could bring to them. There was not much time to schedule and train the stakeholders as well to bring them up to speed. As a result, the stakeholders were unable to align on a Vision. The Vendor was asked to provide explosion, toxic and fire contour maps for specific scenarios and locations that were almost 'fixed' on the site plan at that point in time. The resultant contour maps showed significant consequence on nearby facility with over 700+ people. It was also noted that the plant area consuming the most hazardous substance would impact the nearby facilities significantly from consequence perspective. It should be noted that the result was only based on a selected few scenarios. By the time the project reached the "Execution" phase, more scenarios were added as part of the scope. But it was already too late to incorporate most of the recommended changes resulting from CA/QRA work. The project went through several area by area risk reduction workshops only to accept (by the company executives per organization risk matrix) the aggregate residual risk level which was also found to be not within the organization risk threshold.

2.3.3 Importance of Starting Early

Figure 3: Impact of variable based on project time [5]

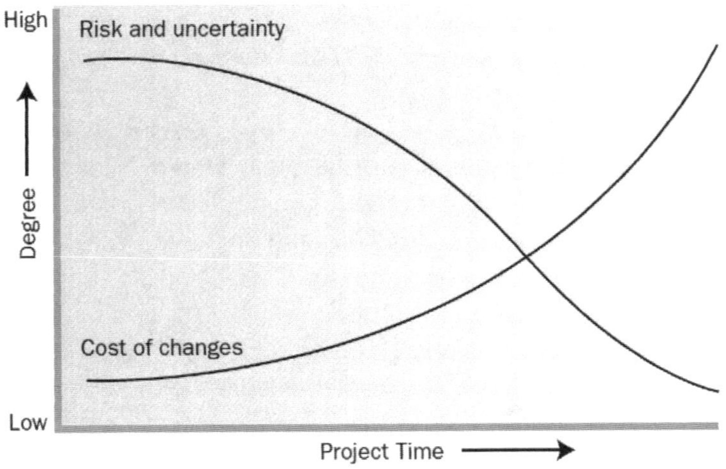

The above two examples highlight the importance of starting CA/QRA work early enough to realize and maximize the value CA/QRA brings to a Mega-Project. The high-level view provides common frame of reference for comparing projects for CA/QRA initiative, even if the projects were dissimilar. Change driven methodology allows for applying the known information at early stage in a way that supports laying out plant facilities to reduce risk and thus maximize potential value. While these characteristics remain present to some extent in almost all project life cycles, they are not always present to the same degree [5]. Change Driven framework, in particular, keeps stakeholder influences higher and the costs of changes lower especially in the early stages of the life cycle than in predictive/plan life cycles. The Figure 3 highlights a well-known fact of project management that even though risk and uncertainty remains high in the early stages of a project, cost of incorporating changes remains much lower. This translates into a key opportunity for project stakeholders to align on laying out the key areas on a site plan intelligently to reduce risk and maximize value by applying tools such as CA and QRA.

2.4 Key Decision Points

Facility Siting, from a management point of view, is all about bringing the stakeholders closer and find the 'right' spot for the facilities. Finding the 'right' spots requires making timely decisions. The challenge has always been to make timely decisions as the stakeholders more often find themselves forced to make those decisions based on information that are frequently of poor quality and at times uncertain, especially in the early stages of a project. At times, due to the scarcity of information, stakeholders rely on their past experience to make those decisions on plant layout. As a result, depending on the individual experience, two different teams for a project area may come up with two very different plant layout. CA/QRA work, if started early enough, will also depend on input that most often requires embellishment through assumptions to create needed scenarios and models. Even with this limitation, CA/QRA provides a key 'input' in allocating real estate for key facilities on a plant site at early stage of a project lifecycle. With each iteration and deliverable 'Change driven' methodology allows for multiple opportunities to make critical decisions on plant layout over the project life-cycle. It allows the stakeholders to decide when to 'sharpen the pencil' (e.g. PHAST to Computational Fluid Dynamics model), relocate facility or accept the consequence and/or risk, 'pull the plug' on further modeling/refinement and start developing justification, if required, to recommend management to accept/reject the aggregate risk profile.

2.5 Alignment with Process Hazard Analysis (PHA) Planning

CA/QRA results are a useful input to PHA (e.g. HAZOP, LOPA) activities. This is particularly true as projects move forward after establishing the plant layout on a high level and starts to conduct "Define Project" (Figure 1). PHA activities may also result in recommendations (i.e. new scenarios) that goes into the *Product Backlog* and requires project team to develop new models or refine old ones. Mega-projects have multiple sub-projects running in parallel most of the time, having an Integrated Schedule for all the reviews is a good strategy so that the right stakeholders are booked for the sessions. The integrated PHA schedule also supports prioritization of *Product Backlog* for CA/QRA activities over the lifecycle of the project.

2.6 Integration Strategy

It is critical for a Mega-Project involving multiple plant areas/silos to understand the BIG picture so that site risk can be dealt with on an aggregate basis. There is strong possibility of one area to impact other plants in the vicinity. It is recommended to review integration aspect and aggregated results on a regular basis. The following Figure 4 shows that the origin of an incident may reside in one particular plant area, while it may impact other areas in the vicinity.

Figure 4: Consequence of an incident in one area impacting multiple areas

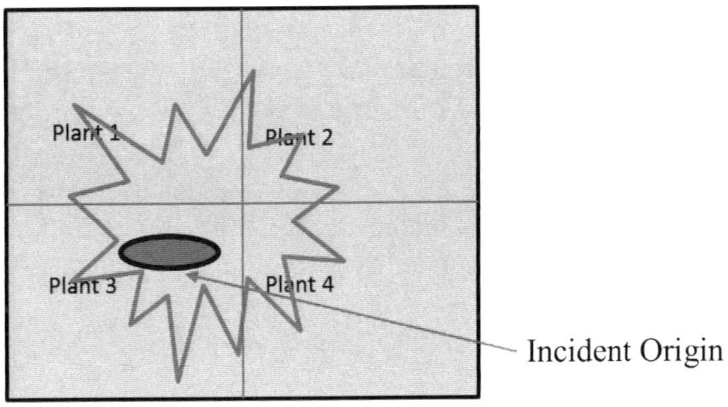

Recommended Strategy:

- Setup Strong Center - Task Force Single Point of Contact (SPOC)
- Collaborate – SPOC will promote Collaboration among Plants/Areas

3. List of Activities (suggested)

3.1 Overall Sequence

The CA/QRA work typically should start with the approved Charter as shown on Figure 1. A project Task Force i.e. SPOC is formed comprising of Project Engineer, Operations rep, Process rep and CA/QRA Technical Expert. The SPOC is ultimately accountable to produce the CA/QRA deliverable studies but also acts to support the design development and site plan finalization efforts in the early stages of the project.

The team collects 'input' such as Preliminary site Layout and design information, works with area process and operations team to identify scenarios, develop assumptions for known-unknowns and unknown-unknowns, develop modeling 'Product backlog', prioritize backlog based on PHA schedule and produce modeling output (i.e. deliverable) for stakeholder use. The stakeholder usually has the option of accepting or rejecting consequence. If the result of an iteration is acceptable within area causing the incident – communicate to outside stakeholders to ensure there is no cross boundary impact. If the result of the iteration is NOT acceptable to outside of the area – use the result as input to PHA for the areas to mitigate. Collaboration recommended between origin area and impacted area/s to identify appropriate mitigations. There is also a possibility that the consequence resulting from the iteration is not acceptable to the area causing the incident, in that case the area will initiate a PHA to mitigate the consequence. As part of developing mitigations, consider the following:

- Develop QRA if appropriate
- Perform Building Assessments for Critical Buildings
- Apply Cost-Benefit analysis to optimize risks/opportunities

- Determine Safe Haven locations and their characteristics

If Mitigation is not possible from Business perspective – perform As Low As Reasonably Practicable (ALARP) justification for Senior Management to formally accept the posed risk

- To apply mitigations on a site wide basis, review Milestone deliverables by taking on a BIG picture perspective
- Future iterations may enhance those deliverables or create new ones as new information is available. Each iteration incrementally builds the deliverables to produce the Mitigated CA/QRA report to support Facility Siting for the stage

The overall effort is expected to produce a maximum of two single reports for the overall site:

1. Mitigated version of QRA Report
2. and Consolidated CA Report

It is recommended to develop as-built, pre-start up CA/QRA reports prior to start-up. Table 1 and 2 outlines typical activities performed by SPOC and plant areas respectively once key process areas are established/finalized on Site Plan. Please note that the activities are project dependent and may warrant modifications to suit specific situation.

Table 1. SPOC Sequence of Activities to conduct Facility Siting

Initiative	Key Steps
Setup Facility Siting Team	Sponsor to support to setup a team (i.e. Project Engineer, Technical Expert, Operation, Process representative) to act as SPOC
Integration	Provide functional (technical and management) guidance to Areas
	Assess cross boundary impact/opportunity
	Communicate incremental outputs after each Area PHA/Risk Reduction
IFA versions (prior to "Define Project" stage)	Consolidated CA Report
	Mitigated and Integrated QRA Report
IFD CA/QRA (prior to "Execution" stage)	Re-validate IFA CA/QRA post PHA and vendor PHA
As-built versions	Post-oil version

3.2 Typical Area Activities

Table 2. Area (Typical) Sequence of Activities to conduct Facility Siting

Initiative	Key Steps
Plant 1 (Typical)	Perform Modeling and Calculations
	Conduct Risk Reduction Workshop (if appropriate) – Apply mitigations
	Incremental output for use based on reduced risk level
	Assess Buildings/facilities
	Identify Safe Haven/shelter Locations (if required)
	Estimate Area Project Scope impact
	Perform area scope updates/changes e.g. passive fire protection requirement
	Carry out QRA (if appropriate)
	Update Product Backlog
	Develop/update contract to support area specific activities

4. Pitfalls to Avoid

The following observations were noted by the Author during his time to manage Facility Siting scope:

1. Facility Siting does not replace good and common sense engineering practice e.g. locating a plant that uses high volume of hazardous material will significantly increase risks to other plants/areas (e.g. camp) within the vicinity
2. Evaluate effectiveness of 'Cost Reduction' philosophies as opposed to added risk e.g. different modularization techniques may decrease the plant footprint by reducing pipe lengths but makes the plant more compact and likely to increase severity of explosion consequence
3. Avoid developing models with hope to realize risk reduction
4. Sharpening the pencil for models most likely may not reduce risk
5. Avoid jumping into CFD or other similar methods without evaluating cost-benefit for undertaking such effort
6. Utilization of QRA results is much more controversial than the methodology. Watson [6] has suggested that QRA should be considered as an argument, rather than a declaration of truth. In this view – focus on accuracy over precision
7. CA/QRA Task Force should be part of Management of Change (MOC) process
8. Evaluate carefully before locating facilities e.g. tank farms close to main entrance/egress as part of Facility Siting

9. Keep the focus on VISION for the overall effort to avoid scope creep
10. Note the difference between Facility Siting as a Deliverable and tool for design-development activities

5. Conclusion

This paper presented Adaptive/Change driven framework to manage initiatives like facility siting/CA/QRA as part of Mega-Projects. Change driven framework does not mean "lack of planning or no planning". As demonstrated in the paper, this methodology starts with a vision and then develops the ultimate product both iteratively and incrementally when it is possible to define small incremental improvements that will deliver value to stakeholders. This allows making key decisions on laying out the facilities in early stages of a project rather than waiting for design to develop to a significant extent. Often times, project team wait too long for design to mature enough before applying CA/QRA to validate consequence impact and/or risk level. Most often, especially for Mega Projects, this leads to delays in executing the CA/QRA work properly at the right time to influence site layout. This delay in execution leads to scenarios when scope changes cannot be incorporated anymore. This translates into value erosion as project was unable to reap the full benefit of Facility Siting. The management framework recommended in this paper i.e. Change driven/adaptive framework allows for the stakeholders to start early in the project lifecycle to apply CA/QRA in laying out the plant facilities. The key is to start EARLY but START with the END in mind. Alignment among stakeholder on a VISION up-front is also important as this is expected to save life while striking a balance between business needs and climbing the Risk Mountain. Vision also allows avoiding scope creep while applying Adaptive/Change driven approach. It is highly recommended that stakeholders focus more on accuracy over precision. In the early stages of project lifecycle, quality information is scarce. Change driven/Adaptive framework gives you as the project stakeholder the opportunity to use "what you know" intelligently to ensure "what you don't know" falls into the right place.

6. References

[1] M. Mannan, A. Chowdhury and O. Reyes-Valdes, "A portrait of process safety: From its start to present day," College Station, 2012.

[2] M. Mannan, Lees' Loss Prevention in the Process Industries, Volumes 1–3 (3rd Edition), , 2005, Elsevier, 2005.

[3] K. McAndrews, "Consequences of Macondo: A Summary of Recently Proposed and Enacted Changes to US Offshore Drilling Safety and Environmental Regulation," in *Society of Petroleum Engineers, Americas E&P Health, Safety, Security and Environmental Conference*, 2012.

[4] E. W. Merrow, Industrial Megaprojects Concepts, Strategies and Practices for Success, New Jersey.: John Wiley & Sons, Inc., 2011.

[5] A Guide to the Project Management Body of Knowledge (PMBOK® Guide) – Fifth Edition, Newtown Square: Project Management Institute, Inc., 2013.

[6] S. Watson, "The Meaning of Probability in Probabilistic Risk Assessment," *Reliability Engineering and System Safety,* vol. 45, pp. 261-269, 1994.

348480 Global Energy and Transportation and Mobil Oil

Abstract:

Biofuels, Computational Methods, Engineering Economics, Polymers, Thermodynamics. Polymers-plastic, Thermodynamics- oil, Engineering Economics – the money it takes to finish the method of Global Engineering Research, Computational Methods – numerical input of analytical data, Biofuels – the transport fluids that utilized in more than one method from industrial usage to everyday living which enters the atmosphere. Exxon has completed a $600 million dollar deal for production of algae biofuel, amongst Mobil Oil has performed major research Global Energy and the effects of oil. Reading a most recent article on Mobil website.

348748 Characterization of Iron Phthalocyanine As the Active Material for Lithium Batteries

Abstract:

Results on the determination of the lithium-ion diffusion coefficient in the iron phthalocyanine (FePc) active material as a function of intercalation level, using a modified electrochemical impedance spectroscopy technique, will be presented. Also, developed thermodynamic functions such as the Gibbs free energy, entropy, and enthalpy changes for the lithium insertion reaction with FePc, derived from cell open-circuit voltage measurements as a function of intercalation level, are presented. Also presented are estimates of the maximum amount of lithium that can be intercalated into the FePc based on constant volume and temperature conditions.

349086 Design of a Free-Fall Reactor for Fast Pyrolysis of Waste Plastics

Abstract:

Thermal pyrolysis has earned significant reputation in the recovery of valuable chemicals, especially petrochemical based fuels from various materials, including biomass and plastics. Of many available technologies, *fast* pyrolysis was preferred in his study owing to its minimal heat and mass transfer limitations and better control on process conditions. The focus of this study is towards efficient conversion of waste plastics into gaseous products consisting primarily of alkanes, alkenes and aromatics. Lab scale experiments were performed using a micro reactor to determine favorable conditions, including reaction temperature and particle size that would result in maximum yield of ethylene. Using these results, a simple industrial size radiatively heated free-fall reactor was designed for the purpose of pyrolyzing waste plastics. The goal is to design a *fast* pyrolysis reactor that improves upon inefficiencies in traditional reactors while resulting in high yield of ethylene.

349732 Biosorptive Dehydration of Ethanol/Water Azeotropes Using Compound Starch-Based Adsorbent

Abstract:

Abstract: Recently, the interest in the production of bioethanol has greatly increased as a result of the increasing fossil fuel prices and associated environmental issues. Bioethanol is made from either grains or biomass and the broth contains 6 to 12 percent ethanol. In addition, water removal represents a serious problem as ethanol form azeotrope with water at atmospheric pressure. Distillation to water-free ethanol consumes 50 to 80 percent of energy used in the fermentation ethanol manufacturing process. Various techniques have been developed to break the azeotropic point, such as salting-out method, azeotropic distillation, extractive distillation, reactive distillation, adsorptive distillation and pervaporation. Among the methods, the adsorptive option is the most attractive, especially the biosorption, which has distinct advantages over conventional methods as it is environmentally friendly, highly selective, easily available, easy to operate, cost effective and reusable in repeated cycles in the treatment of aqueous organics.

In this work, a specially formulated compound starch-based adsorbent (CSA), which consists of corn, sweet potato and foaming agent, was developed. As an exploratory study, this research was focused on the investigation of CSA for biosorptive dehydration of aqueous ethanol and optimization of conditions for its maximum adsorption under various conditions of vapor feed flow rate, bed temperature and bed height through an orthogonal design method. Furthermore, to investigate the adsorption mechanism, dynamic saturated adsorbance of CSA was studied and characterization of the adsorbent before and after adsorption, as well as after regeneration, was performed.

In the first step, the net retention time and separation factor of ethanol and water were obtained using inverse gas chromatography (IGC), the results indicated that the net retention time of water on CSA is significantly longer than that of ethanol which attested the feasibility of this adsorptive separation purpose and low temperature could be propitious to the process which was consistent with the traditional theory of adsorption.

Then, through a batch adsorption experiment, the optimum adsorption condition was determined under vapor feed flow rate, bed temperature and bed height by orthogonal design method, which indicated that the optimization condition for ethanol was at vapor feed flow rate of 1.1g/min, bed temperature of 83 and bed height of 25cm. The packing density was 445kg/m^3. The residence time is determined by vapor feed flow rate and bed height. The residence time is more influenced than bed temperature. The longer the residence time, the more the output is. But it cannot be ignored that the adsorption selectivity may decreases. Low bed temperature is benefited to the adsorption process which agrees with the result of IGC. Five cycles experiments of adsorption-regeneration were carried out at the optimum adsorption and regeneration condition, the results showed that there is practically no change in the regeneration efficiency of CSA after the cyclic process adsorption–regeneration. It is due to the properly chosen regeneration conditions assuring complete desorption of the water adsorbed without changes in the CSA structure. This unchanged adsorption capacity after a number of cycles demonstrates reusability of the adsorbent.

To study the adsorption mechanism, experiment was conducted to study the dynamic saturated adsorbance of CSA. The water adsorption capacity of CSA has been found to be similar to some molecular sieves, which is about 0.15g/g. Besides, field emission scanning electron microscopy (FESEM) and mercury porosimetry (MP) were applied to investigate the change of surface morphology and microstructure of CSA before and after the adsorption, as well as after regeneration, as shown in Fig.1 and Fig.2. The FESEM images showed that compared with the particles before the adsorption, they became plump and the void spaces became smaller after adsorption, which indicated that the CSA was saturated with water. However, the CSA particles after regeneration became wizened like the nature particles, which demonstrated that almost all the water had been desorbed from CSA. The pore size distribution (PSD) indicated that compared with the PSD curve before adsorption, the peak at 6000nm disappeared and the peak at 17000nm became remarkably smaller after adsorption, which manifested that CSA is mainly composed of macropores which play a major role in the adsorption. However, the peaks at 6000 and 17000nm reappeared after regeneration, and the curve looked exactly the same with that before adsorption, which further demonstrated that CSA after regeneration can be totally regenerated and recovered to the original structure.

These results indicated that the starch-based adsorbent can be considered as an effective and low-cost bio-sorbent to dehydrate alcohol.

349781 Purification of 2-Amino-1-Phenylethanol Enantiomers By a New Technique Combining Distillation and Crystallization

Abstract:

Stripping crystallization (SC) is a new separation technique which combines distillation and crystallization. Basically, SC is operated at a triple-point condition, in which the liquid is simultaneously vaporized and crystallized due to the three-phase equilibrium. In essence, the process is continued until the liquid phase is completely eliminated and only the pure solid crystals remain in the feed. Thus, crystal washing is not required since only pure solid crystals remain in the feed and no impurities are adhered on the crystal surfaces at the end of SC. In the previous research, SC has been successfully applied to separate p-xylene from the mixed xylenes by Shiau et al. (2005, 2008).

Phenylethanolamine, also known as 2-amino-1-phenylethanol (2-APE), is an important basic skeleton in many adrenergic drugs and is often used as ingredients of cold medicines and anorectics. As the biological activity and pharmacokinetics of the enantiomers of a chiral drug can be different, the objective of this research is to apply SC to purify R-2-APE from the chiral 2-APE mixture. The experiments show that no crystallization occurs during SC without seeding due to subcooling of the liquid enantiomers. On the other hand, seeding can eliminate subcooling during SC, leading to the R-2-APE crystals gradually formed preferentially on the seeds. Thus, seeding with magnetic stirring can be successfully employed to purify R-2-APE from the 2-APE enantiomers.

1. Shiau, L. D., Wen, C. C. and Lin, B. S. Separation and purification of p-xylene from the mixture of m-xylene and p-xylene by distillative freezing, *Ind. Eng. Chem. Res*, **44(7)**, 2258-2265, 2005.

2. Shiau, L. D., Wen, C. C. and Lin, B. S. Separation of p-xylene from the multicomponent xylene system by stripping crystallization, *AIChE Journal*, **54(1)**, 337-342, 2008.

350841 Fabrication of Low Cost Insulating Material from Kaolin Clay for Construction Purposes

Abstract:

A new low cost, thermal insulator, construction mineral polymeric material was synthesized from local Saudi Arabia resources using Inorganic Mineral Polymerization technique (MiP). In this technique, the mineral polymer is produced from the following components: clay mineral rich in koalinite, sand, sodium hydroxide solution, and water. Inorganic polymerization experiments are conducted by mixing known amounts of clay minerals and sand in a variable speed stainless-steel mixer for some period of times until a homogenous solid mixture is obtained. Sodium hydroxide solution is added to the mixture and the polymerization reaction takes place at different mixing speeds and times. The produced mixture is divided into a number of samples of known weights. Each sample is moulded in a stainless-steel cylinder at a medium pressure. Then the moulded samples are cured in an oven for 24 hours. The curing temperature was 80°C. The mineral polymerization conditions such as the ratio of clay to sand filler, particle size of the clay, particle size of the sand, concentration and amount of sodium hydroxide, speed and time of mixing were optimized to obtain stable and hard mineral polymer. Compressive strength for different specimens of fabricated mineral polymers was tested under dry and wet conditions. Values of compressive strength with values ranging from 11.5 MPa to 27.9 MPa were obtained.

351141 Thermodynamic Study of Binary PAH (Anthracene + Phenanthrene) Solid Mixtures

Abstract:

Polycyclic aromatic hydrocarbons (PAHs) are common components of many materials, such as petroleum and various types of tars. They are generally present in mixtures, occurring both naturally and as byproducts of fuel processing operations. It is important to understand the thermodynamic properties of such mixtures in order to better understand and predict their behavior (i.e., fate and transport) in the environment and in industrial operations. The Knudsen effusion technique was combined with gas chromatography and differential scanning calorimetry to measure the vapor pressure, composition, and phase behavior of binary anthracene and phenanthrene mixtures at near ambient temperatures. Mixtures of anthracene and phenanthrene exhibited non-ideal mixture behavior: In mixtures with less than 80% phenanthrene, this more volatile component preferentially sublimed and the vapor pressures decreased as the mixtures became enriched in anthracene. The vapor pressure data do not follow Raoult's law (that is, ideal mixture) predictions. Mixtures with high phenanthrene content (> 80%) showed azeotrope like behavior, in that anthracene and phenanthrene sublimed at nearly constant vapor pressure and molecular ratio. These results confirm that the behavior of few-component PAH mixtures does not follow ideal mixture models. Hence, as tars and petroleum (and potentially other organic compound-containing mixtures) weather to few component mixtures over time, complicated, non-ideal phase behavior might be observed.

351160 How Confident Are You That Your Major Accident Risks Are Under Control?

Abstract:

Major accidents in the process industries including those at Texas City , Buncefield UK and in the Gulf of Mexico have highlighted the critical role played by senior managers and executives. Effective leadership is essential if companies are to develop a positive safety culture that remains constantly vigilant towards the risk of a major process safety incident. In the UK the HSE has made leadership a key priority for the high hazard industries and there is a cross-industry initiative to develop and deliver training focussed towards board members responsible for process safety performance.

This paper explores the leadership failings that contributed to recent major accidents, and the cost to both the company reputation and the individuals involved. Key principles are discussed that can help improve leadership in the major hazard industries. These include; attitudes and decisions of senior manager affecting the investment strategy and the safety culture of the organisation, reinforcing the importance of safety by personal example, thorough understanding of major accident hazards and key risk control systems, investigating process safety incident and near misses to find the underlying causes, developing world class safety management systems, and identifying weaknesses in these systems using targeted performance indicators.

The author bases this paper on the development of a training standard for 'Process Safety Leadership' involving a cross-industry 'expert panel'. Training of senior executives in compliance with this standard has been carried out from late 2011 and feedback from these courses and the key requirements for effective leadership will be discussed.

351236 Sil Determination: Shortcomings with the Use of LOPA

Abstract:

Proposed Topic Area: Managing Risks

For a number of years, industry has been trying to apply the principles outlined in IEC 61508. In the process sector, this has resulted in a focus on the requirements of IEC 61511 (2003) and ANSI/ISA 84.00.01 (2004). These are risk based standards. They involve the identification of potential hazardous events, the causes that could lead to each event and the necessary risk reduction for each event. Where the risk reduction involves an instrumented function (e.g. a trip or an interlock), the required performance of the instrumented function can described in terms of the average probability of failure on demand. In the standards, the performance for such instrumented functions is assigned to one of four Safety Integrity Levels. The assessment of what performance level is needed in respect to a specific hazardous event is referred to as SIL Determination. In the field of SIL Determination, IEC 61511 provides examples of a number of different approaches – these are shown in IEC 61511 Part 3. These include Event Trees, Risk Graphs, Safety Matrices, Fault Tree Analysis and Layer of Protection Analysis (LOPA).

In the first few years, following the publication of IEC 61508 Edition 1 in 2000, industry generally embraced the use of Risk Graphs as the method of choice. However, over the subsequent few years, many companies using this method found that it was (a) giving higher safety integrity level targets than really needed and (b) was less flexible than other methods and consequently more difficult to use in many real situations. This has resulted in a general move to other methods and in particular to the use of Layer of Protection Analysis. Indeed, LOPA may now be the predominant "Method of Choice" for SIL Determination. As a method, LOPA gives the impression of simplicity and appears quite straightforward. It almost invites anyone with a minimum of understanding of the method to be a practitioner. However, there are a number of shortcomings in the use of the LOPA method that have become evident through its use across the globe.

This paper will review the application of LOPA and provide insights into what may be seen as the proper use of the method and its core principles. It will also highlight various shortcomings in use that have been found during expert reviews carried out of a number of existing studies. The paper will be of benefit both to those who are new to the LOPA method, but also of value to those who have been using the method for some time.

Shale Oil and Gas – Wellhead to Fuel

Dr. Lawrence Kremer - Baker Hughes
AIChE - Hilton New Orleans Riverside
April 1, 2014

14AIChE
Spring Meeting, New Orleans, LA

BAKER HUGHES

Outline Unconventional Shale Oil

- Properties of shale oils
- Problems
  - Transportation
  - Refining

14AIChE
Spring Meeting, New Orleans, LA

BAKER HUGHES

General Characteristics of Shale Oils

- API gravity typically varies 30-55+
- Very paraffinic
- Low sulfur
- Can have hydrogen sulfide and mercaptans
- Low TAN
- Very low asphaltenes
- Light crude with very little residual material
- Many samples show great quality variation

Factor	Bakken	WCS
API	39.7	21.6
S%	0.25	3.44
Light ends	2.6%	
LSR	3.5%	3%
Light Nap	12.4	3.0
Heavy Nap	9.7	5.0
Jet	10.5	10.0
LFO	27.6	12.0
Gasoil	27.2	37.0
VacTar	6.7	29.8

14AIChE Spring Meeting, New Orleans, LA

BAKER HUGHES

Shale Oil Property Comparison

Parameter	Eagle Ford	Bakken
API	52	40.8
TAN (g KOH/g)	< 0.05	0.09
Sulfur (wt %)	<0.2	0.304
Asphaltene (wt %)	0.1	0.41
Resin (wt %)	1.6	4.95
Filterable Solids (PTB)	225	76

14AIChE Spring Meeting, New Orleans, LA

BAKER HUGHES

Utica Shale Crude - Three Phase

14 AIChE
Spring Meeting, New Orleans, LA

BAKER HUGHES

Eagle Ford Crude Samples

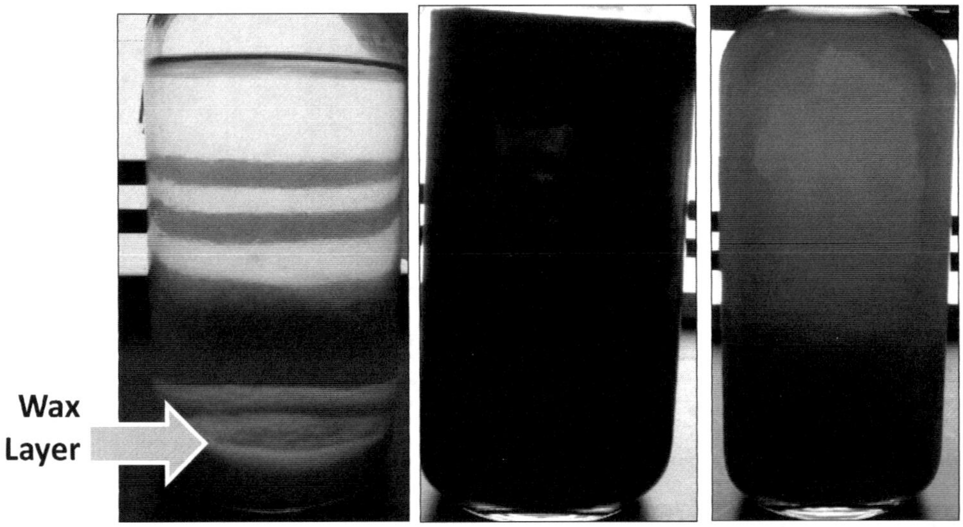

14 AIChE
Spring Meeting, New Orleans, LA

BAKER HUGHES

Analytical Breakdown

Value	Yellow	Red	Black
API	55.0	44.6	52.3
TAN	<0.05	0.07	<0.05
Sulfur	<0.20%	<0.2%	<0.2%
Na	1.0 ppm	1.6 ppm	1.6 ppm
K	0.3 ppm	0.4 ppm	0.5 ppm
Mg	3.4 ppm	2.9 ppm	3.0 ppm
Ca	2.6 ppm	2.8 ppm	3.8 ppm
Asphaltene	0%	0%	0.1%
Resin	0.5%	3.2%	1.6%
Filterable solids	176 PTB	295 PTB	225 PTB
P	5.3 ppm	4.6 ppm	5.0 ppm

14AIChE Spring Meeting, New Orleans, LA

BAKER HUGHES

Micrograph of Sludge

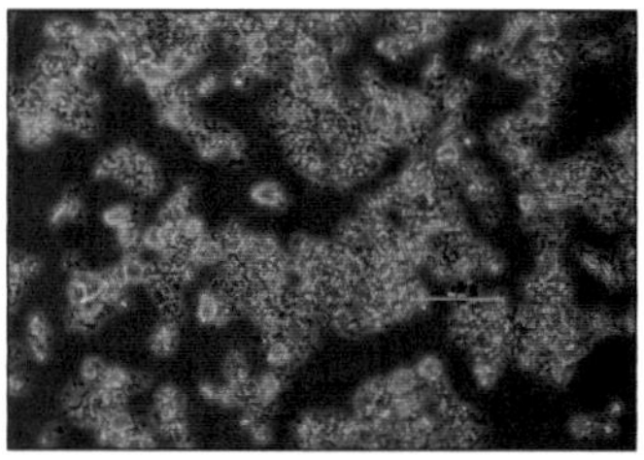

100 X

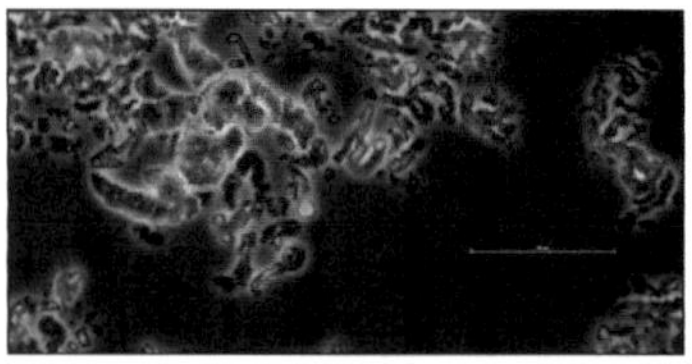

400 X

14AIChE Spring Meeting, New Orleans, LA

BAKER HUGHES

Paraffin Analysis of Sample

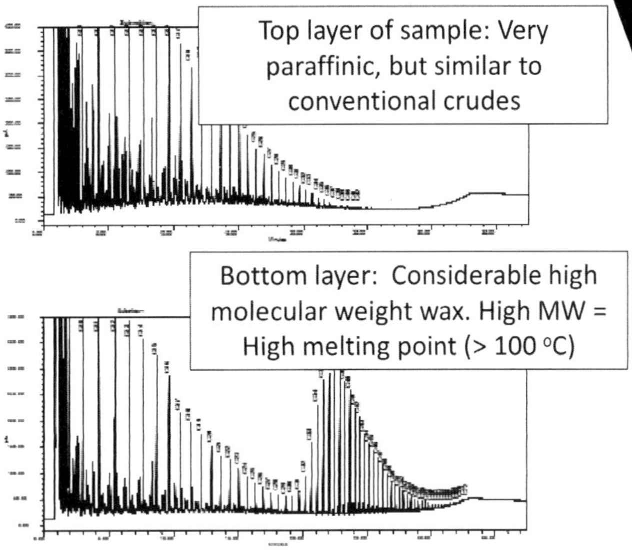

Transport of Shale Production

- Infrastructure does not exist to meet needs
  - Most shale oil moved by rail or trucks
  - Limited pipelines expanding to meet transportation needs
- Unit trains to East Coast, West Coast, and Gulf Coast
- Problems:
  - Tankers and barges with heavy wax deposits ROB
  - Tight emulsions that look like oil or white condensate but high BS&W
  - Cars not uniformly treated for H_2S
  - Corrosion in rail cars

Viscosity of ROB Before Treatment

ROB from untreated compartment

14 AIChE
Spring Meeting, New Orleans, LA

BAKER HUGHES

Eagle Ford Shale Oil Wax Dispersion Testing

Several products were effective in dispersing the paraffin wax in an Eagle Ford shale oil sample at ambient temperature for over a month.

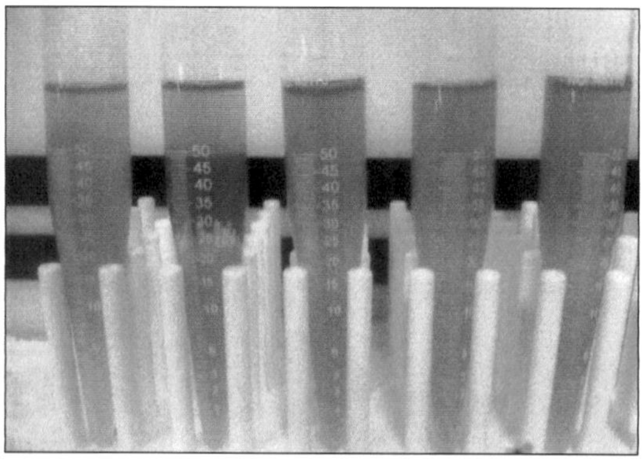

14 AIChE
Spring Meeting, New Orleans, LA

BAKER HUGHES

Product 'A' Effect on Wax Deposition (Utica)

Blank **200 ppm** **400 ppm**

14AIChE® Spring Meeting, New Orleans, LA

BAKER HUGHES

Refinery Issues

- Crude Blending
- Fouling
- Corrosion
- Product quality

14AIChE® Spring Meeting, New Orleans, LA

BAKER HUGHES

Crude Unit Fouling

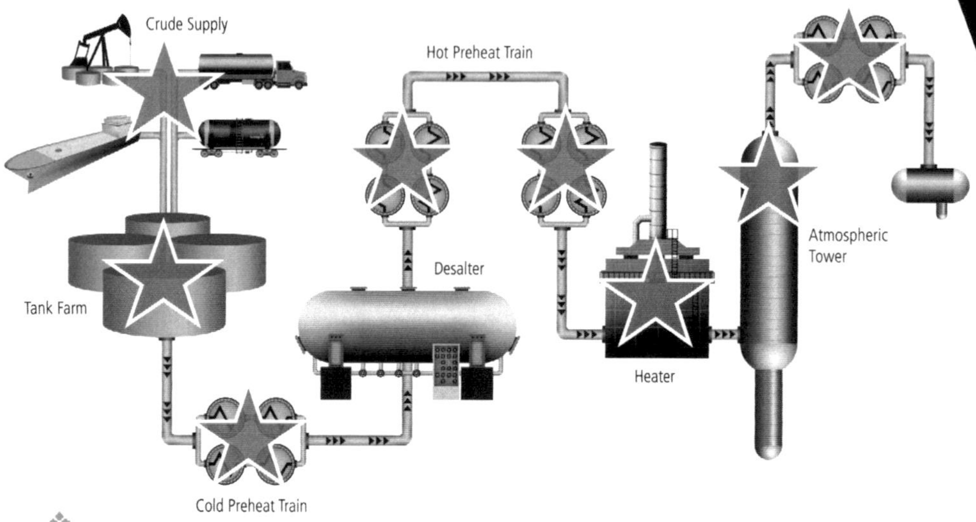

14 AIChE Spring Meeting, New Orleans, LA

BAKER HUGHES

Asphaltene Precipitation

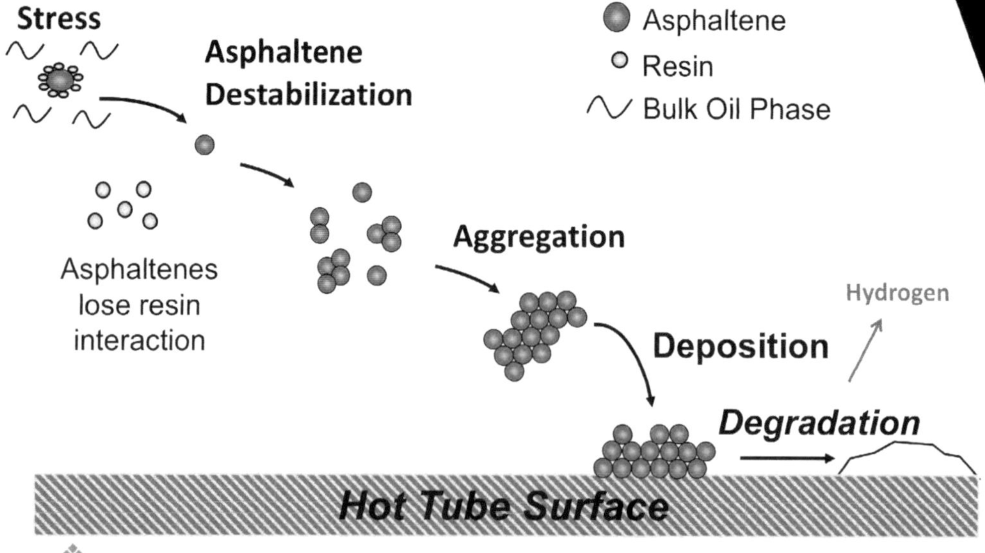

14 AIChE Spring Meeting, New Orleans, LA

BAKER HUGHES

Asphaltene Stability Index Test

Field ASIT services™ technology measures the onset of the flocculation of the asphaltenes with high accuracy by inducing the asphaltene precipitation via titration with a paraffinic solvent.

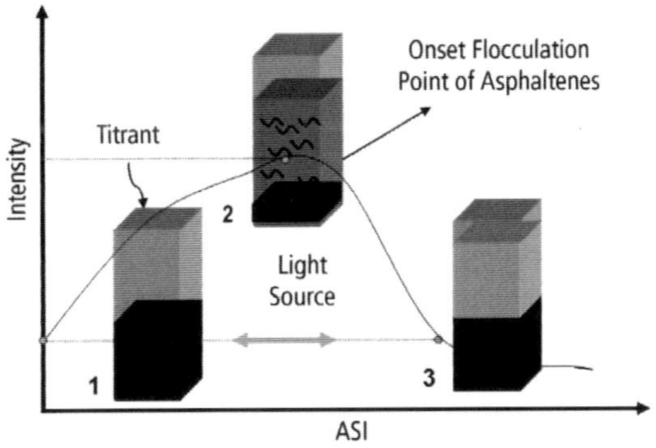

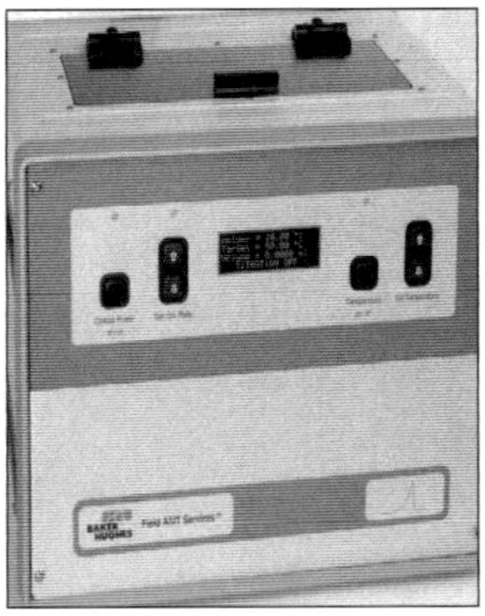

14 AIChE® Spring Meeting, New Orleans, LA

Field ASIT services is a trademark of Baker Hughes Incorporated

BAKER HUGHES

ASI Shows Eagle Ford Can Destabilize Crude

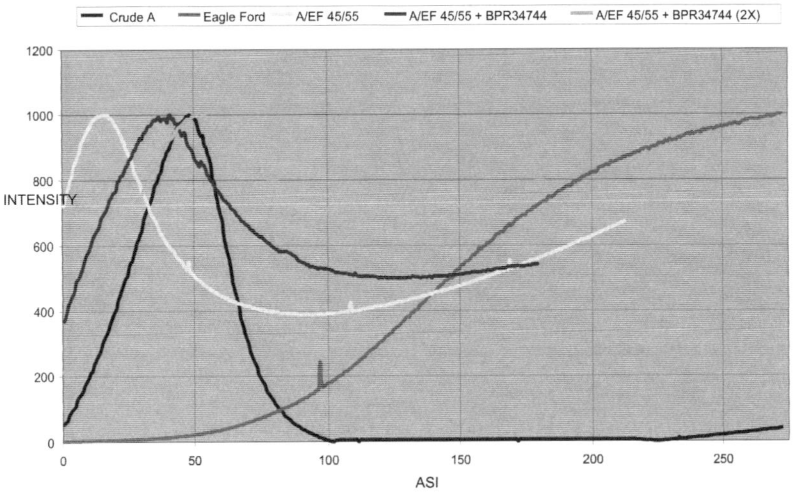

14 AIChE® Spring Meeting, New Orleans, LA

BAKER HUGHES

Tower Tray Fouling - Phosphorus

- Seen by refineries running Bakken, Eagle Ford, and WTI
- Operationally manifested as increasing tower dP
  - Typically in the kero/jet zone
- Tower inspection shows deposits on trays in kero/jet zone
  - Some deposition may also show a few tray levels above and/or below kero/jet zone
  - Deposit scale typically tightly bonded to tray surface restricting flow and valve function

Above kero/jet zone

In kero/jet zone

Below kero/jet zone

14 AIChE
Spring Meeting, New Orleans, LA

BAKER HUGHES

Shale Oil Impact – MEA Contamination

Corrosion and Fouling

- Shale oils transported by truck, rail, barge.
- Many shipments are treated with H_2S scavengers:

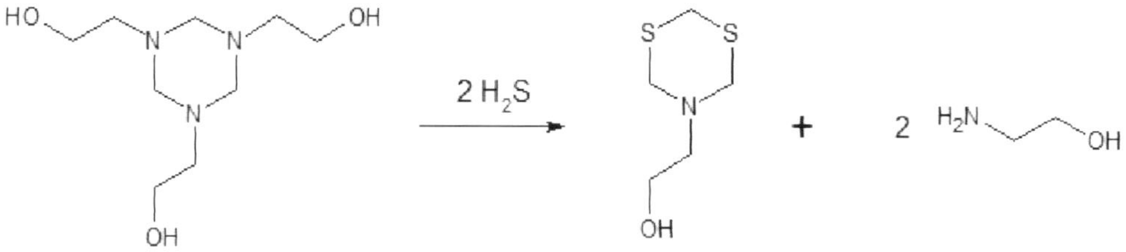

- Potential Impacts:
  - H_2S scavengers leave byproduct amine in crude oil (MEA, MA).
  - Tramp amines distill to OH system, create amine-HCl salts:
  - Corrosive to system metallurgy.

14 AIChE
Spring Meeting, New Orleans, LA

BAKER HUGHES

Impact of Tramp MEA on OH Water

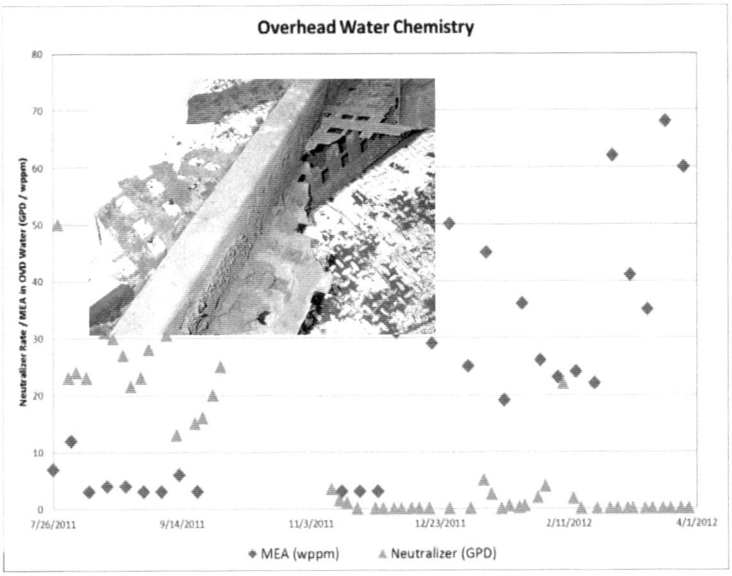

14 AIChE
Spring Meeting, New Orleans, LA

BAKER HUGHES

Salt Risk w/MEA Concentration

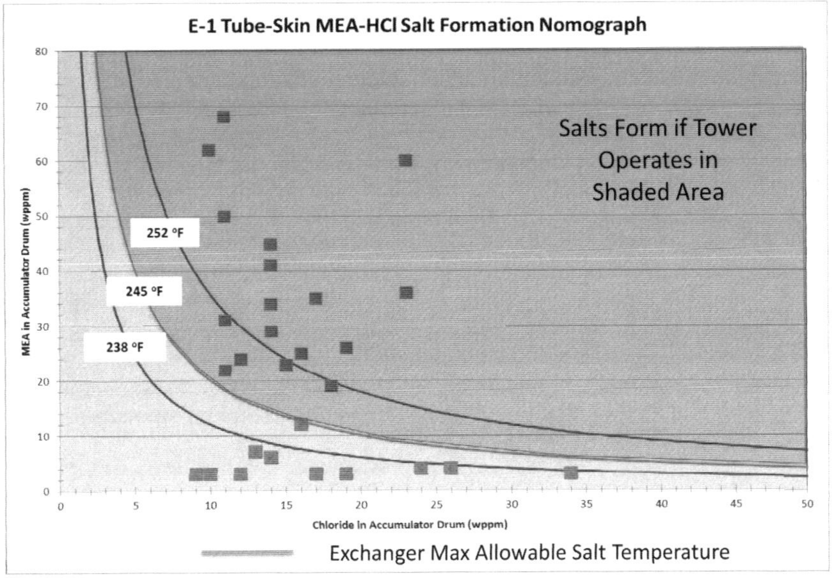

14 AIChE
Spring Meeting, New Orleans, LA

BAKER HUGHES

Shale Oil – Refinery Process Concerns and Monitoring

BAKER HUGHES

CHARACTERISTIC OF SHALE OIL	CONCERN IN REFINERY	PROCESS DATA	SAMPLES AND ANALYSES: Samples	SAMPLES AND ANALYSES: Analysis	OUTPUT DATA	FREQUENCY	BHI PROGRAMS
Crystalline Wax	Tank bottom sludge	• Tank Sludge Depth	• Tank bottom	• Visual sludge • GC analysis	• Tank Sludge Depth • Crystal Wax Presence	• Monthly	• XERIC™ Pretreatment
	Cold Preheat Fouling	• Desalter Temperature			• Temp vs. Time	• Update 2x/mo.	• LIFESPAN™ Wax Dispersants
	Desalter emulsions	• Emulsion layer growth	• Triline emulsions	• Visual wax • GC analysis	• Crystal Wax Presence	• After Shale Oil begins • When emulsion grows	• XERIC Demulsifiers
Asphaltene Stability in Blends	Desalter emulsions	• Emulsion layer growth	• Triline emulsions	• Precipitated Asphaltenes	• Precipitated Asphaltenes	• After Shale Oil beings • When emulsion grows	• XERIC Asphaltene Stabilizers • Field ASIT services™
	ColdPreheat Fouling	• Desalter temperature	• Crude blend	• ASIT	• Temp vs. Time • ASIT #	• Pre-Shale Oil • Update 2x/mo.	• LIFESPAN Antifoulants • Field ASIT services
	Hot Preheat Fouling	• Furnace inlet temp • Furnace tube temps • Furnace fuel usage • Preheat DP	• Crude blend	• ASIT	• Temp vs. Time • Fuel use vs. Time • ASIT #	• Pre-Shale Oil • Update 2x/mo.	• LIFESPAN Antifoulants • MILESTONE™ Furnace Fouling Control • Field ASIT services
Paraffinic	Furnace Coking	• Furnace tube temps • Furnace fuel usage	• Crude blend	• HLPS	• Coking tendency	• Pre-Shale Oil • W/Shale Oil	• MILESTONE Furnace Fouling Control
Tramp MEA	Tower/OH corrosion (Early Warning)	• OH Neutralizer rate • EXCAL usage rate	• OH water • Desalter brine	• OH pH • Brine pH	• Early Sign of Amine • Early sign of Amine	• Pre-Shale Oil • W/Shale Oil	• TOPGUARD™ Overhead Corrosion Control • EXCALIBUR™ Contaminant Removal Technology
	Tower/OH corrosion (Quantitative)	• Inputs for CRM	• OH water • Desalter ww • Desalter brine	• IC Amines	• Confirm MEA in Crude • OH MEA ppm • Salt Risk Tower • Salt Risk OH System	• Pre-Shale Oil • W/Shale Oil • Update 2x/mo.	• TOPGUARD Corrosion Control • EXCALIBUR Technology • TOPGUARD Corrosion Risk Monitor (CRM)
Phosphorus	Tower P Fouling	• Tower DP			• Tower DP Increase	• 2x/mo.	• LIFESPAN Antifoulants, Dispersants
			• Crude	• ICP for P	• P in Crude	• 2x/mo.	
			• Kerosene	• Filt Solids • XRF Filter P • ICP for P	• Incr. Kero Solids • P in Solids • P in Kerosene	• 2x/mo.	
			• Kerosene • Diesel	• ASTM D86	• Incr. in Diesel IPB – Kerp EP overlap	• 2x/mo.	

EXCALIBUR, FIELD ASIT SERVICES, LIFESPAN, MILESTONE, TOPGUARD and XERIC are trademarks of Baker Hughes Incorporated.

SAS

Multipoint Flare History

ZEECO

Slide 2

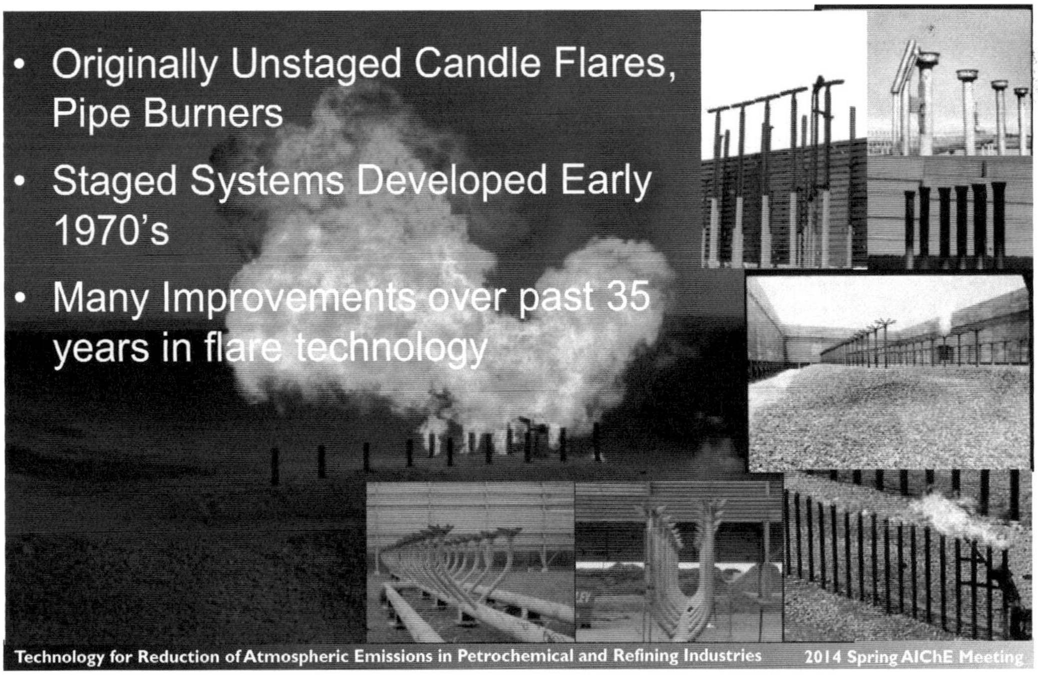

SAS

Staging Allows Large Variation in Flowrates to Maintain Smokeless Operation

ZEECO

Slide 3

Technology for Reduction of Atmospheric Emissions in Petrochemical and Refining Industries 2014 Spring AIChE Meeting

SAS

Large Flare System Design Issues

ZEECO

Slide 4

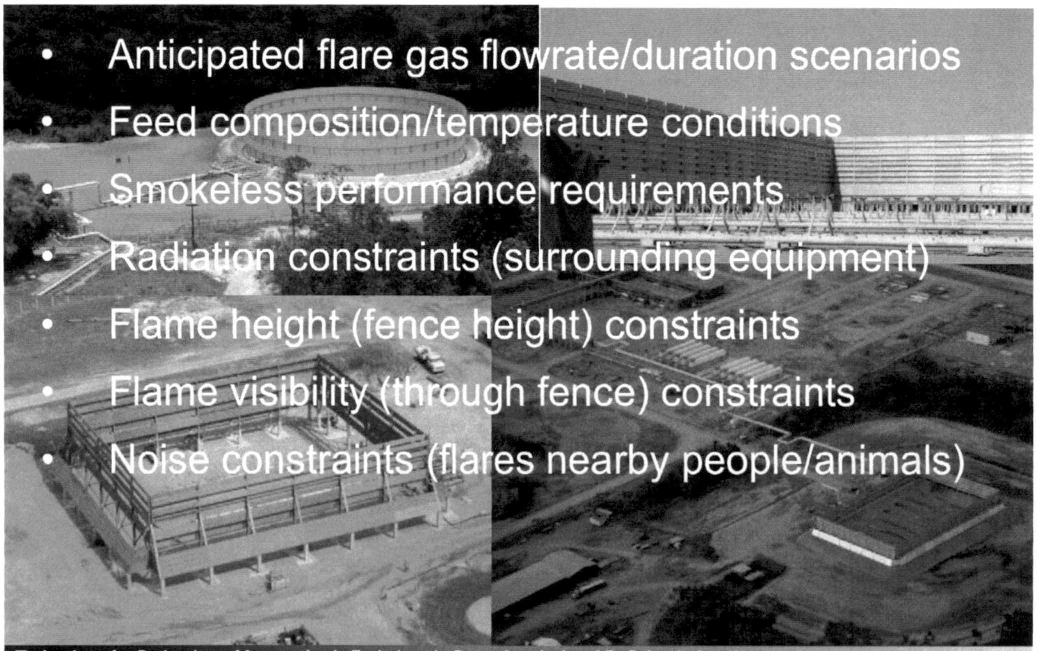

Technology for Reduction of Atmospheric Emissions in Petrochemical and Refining Industries 2014 Spring AIChE Meeting

Burner Tip Design based on Multi-arm/ multi-jet configuration

Slide 5

- Variable Arm Area
- Investment Cast
- Pressure Tested at Factory
- 310 SS Cast Material

Technology for Reduction of Atmospheric Emissions in Petrochemical and Refining Industries 2014 Spring AIChE Meeting

Flare Regulations and Performance Guidelines

Slide 6

- Flaring regulation (API guidelines) focus on utility flares and assisted flares (little attention to multi-tip ground flares)
- Flare Design/Performance guidelines established for general flares – ground flares have additional performance metrics to consider
- Flare performance difficult to quantify (large flame/high radiation flux, elevated operation)
- Multi-Tip ground flares have higher radiation inside large fenced region (larger plumes/difficult to measure emissions!)

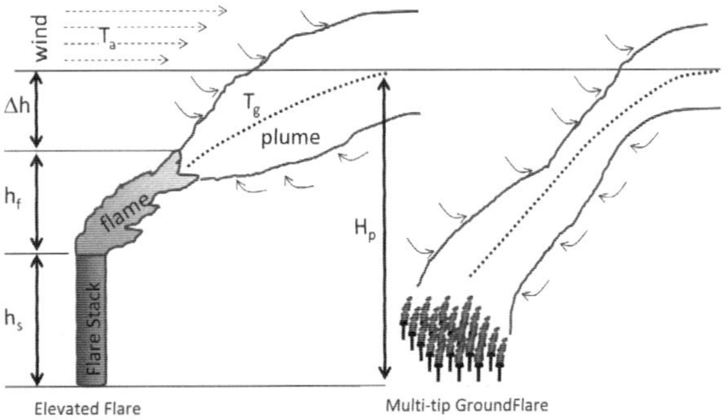

Technology for Reduction of Atmospheric Emissions in Petrochemical and Refining Industries 2014 Spring AIChE Meeting

SAS

Existing Flare Regulations and Guidelines

Slide 7

- 40 CFR 60.18(b) and 63.11(b)
  - Pilot flame required
  - Sufficiently low exit velocity for stable combustion
  - Heating value high enough for flammable operation
  - Smokeless operation requires adequate mixing energy (air)
- 40 CFR 60.11(d), 60.18(d), 60.482-10(e), 63.6(e), 63.11(b)(1), 63.172(e)
  - Adhere to flare's design documents
  - Use Good Air Pollution Control Practice
- New Source Review

Federal requirements for flares are found in the New Source Performance Standards (NSPS) in § 60.18 and National Emission Standards for Hazardous Air Pollutants (NESHAP) in § 63.11. At a minimum, these rules require flares to be:

- Designed and operated with no visible emissions using EPA Method 22 (except for periods not to exceed 5 minutes in 2 hours);
- Operated with a flame present at all times, confirmed by the use of a thermocouple or equivalent device;
- Used only when the net heating value of the gas to be combusted is 300 BTU per standard cubic foot (BTU/scf) or greater (if the flare is steam- or air-assisted), or 200 BTU/scf or greater (if the flare is nonassisted); and
- Designed for and operated with an exit velocity less than 60 feet per second (ft/sec). An exit velocity of greater than 60 ft/sec but less than 400 ft/sec may be used if the net heating value of the gas being combusted is sufficiently high.

Enforcement Alert, "EPA Enforcement Targets Flaring Efficiency Violoations", EPA-325-F-012-002,August (2012).

SAS

Understanding Multi-Tip Ground Flares

Slide 8

- Understanding ground flare performance
  - Where does air come from?
  - How does wind effect flame?
  - How does tip geometry (pressure) affect flame?
- Reacting multi-phase flow in wind
  - Transient flame analysis
  - Near tip mixing
  - 100's of tips in large flare fields
- Model Validation Required
  - Calibrate Smoke and Radiation models

Tip Geometry

SAS INCORPORATED

SAS's C3D Flare Model: Beyond Traditional Flare Analysis

ZEECO

Slide 9

- Extends standard "Steady-State" Flare Analysis:
  - Flame structure interaction with combustion chemistry and radiation flux
  - Fluid interactions with surrounding obstacles
  - Fuel specific soot formation and impact on radiation transfer
- Captures flame dynamics: Momentum vs. Buoyancy, Wind Effect on Flame Structure, Chemical Reactions, Radiation to/from flame
- Assess operating scenarios for specified conditions: wind speed, fuel flow/composition/temperature, impact on/by surroundings
- Key Assumptions:
  - Flame emissivity = F(flare gas composition, soot volume fraction, flame size/shape)
  - Optically thick flames (diffuse radiation) with time dependent view factors
  - Surroundings participate in radiation heat transfer (Re-radiation from ground)
- Overnight analysis on "standard" workstation, animations used to investigate transient flame behavior

Technology for Reduction of Atmospheric Emissions in Petrochemical and Refining Industries 2014 Spring AIChE Meeting

SAS INCORPORATED

Multi-Tip Ground Flares: Modeling Issues

ZEECO

Slide 10

- High tip velocity increases air entrainment
  - Tip design critical to air entrainment
  - Local high velocity can translate into high sound levels
- Assist media not available to increase combustion air
  - smoke below certain tip pressure (D-stage pressure)
- Tip spacing critical
  - Flares must cross light
  - Possible Flame merge lengthens flames
  - Adjacent rows compete for air (longer flames, poor performance)

Technology for Reduction of Atmospheric Emissions in Petrochemical and Refining Industries 2014 Spring AIChE Meeting

SAS

Multipoint Ground Flare Modeling Approach

ZEECO

Slide 11

Single Burner Test/Model
- Detailed mesh
- Radiation/soot tests
- Single flame height/size

Multi-Burner Test/Model
- Radiation/soot tests
- Flame interactions (spacing)
- Multi-Flame shape/size

Model Full Flare Field
- Use Calibrated Models
- Predict Flare Performance (Smoke/Air Demand)
- Predict Radiation Flux

Technology for Reduction of Atmospheric Emissions in Petrochemical and Refining Industries 2014 Spring AIChE Meeting

SAS

Transient Flare Analysis: Multi-step Reaction Mechanism

ZEECO

Slide 12

Incomplete Combustion of Flare Gas:

$$1\text{kg}\ \mathbf{F} + (2.87\text{-}2.6S_1)\ \text{kg}\ \mathbf{O2} \rightarrow S_1\ \text{kg}\ \mathbf{C} + (3.87\text{-}3.6S_1)\ \text{kg}\ \mathbf{PC} + (50\text{-}32S_1)\ \text{MJ}$$

Where: (F); Oxygen (O2); Soot (C); Products of Combustion (PC), S_1 soot parameter based on fuel type

Endothermic fuel pyrolysis (cracking reaction – soot producing):

$$1\text{kg}\ \mathbf{F} + 0.3\ \text{MJ} \rightarrow S_2\ \text{kg}\ \mathbf{C} + (1\text{-}S_2)\ \text{kg IS}$$

Where: Intermediate Species (IS), S_2 is cracking parameter based on fuel type

Soot combustion:

$$1\text{kg}\ \mathbf{C} + 2.6\ \text{kg O2} \rightarrow 3.6\ \text{kg}\ \mathbf{CO2} + 32\ \text{MJ}$$

Where: Carbon Dioxide (CO2) is combustion product

Intermediate Species combustion:

$$1\,\text{kg}\ IS + \frac{2.87 - 2.6S_2}{1 - S_2}\text{kg}\ O_2 \rightarrow \frac{3.87 - 3.6\,S_2}{1\text{-}S_2}\text{kg}\ PC + \frac{50 - 32S_2}{1\text{-}S_2}\ MJ$$

Coefficients selected so complete combustion of carbon soot (C) and intermediate species (IS) produce same species and thermal energy as direct combustion of fuel; Coefficients in terms of mass weight (not moles)

Technology for Reduction of Atmospheric Emissions in Petrochemical and Refining Industries 2014 Spring AIChE Meeting

SAS

Flare Burner Tests

Slide 13

SAS

Single Tip Burning Propane: wind vs. no-wind

Slide 14

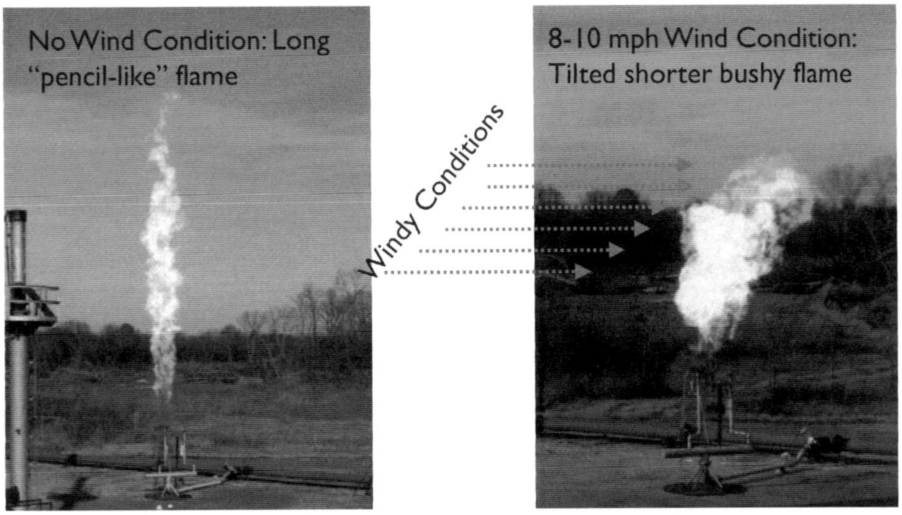

SAS

Single Burner Flare Model

Slide 15

- 6 X 6 X 26 m physical domain
- Flare Tip located 2 m above ground level
- Turbulence and Arrhenius kinetics Included for fuel gas
  - Reaction Parameters adjusted to match observed flame characteristics
  - Soot Yield matched flame height (i.e., soot burnout)
- Flare Movies for no wind, 3m/s (7mph) wind conditions
- Predicted results for Air demand and Radiation loss from flame determined

Technology for Reduction of Atmospheric Emissions in Petrochemical and Refining Industries 2014 Spring AIChE Meeting

SAS

Combustion Models

Slide 16

Propane:

$C_3H_8 + 3.6\ O_2 \rightarrow 3\ CO_2 + 1.6\ H_2O + 0.024$ Soot + 46 MJ/kg propane

Soot + $2.66\ O_2 \rightarrow 3.66\ CO_2$ + 32 MJ/kg Soot

Ethylene:

$C_2H_4 + 0.57\ O_2 \rightarrow 0.93\ C_2H_2 + 0.64\ H_2O$ + 9.4 MJ/kg ethylene

$C_2H_2 + 2.58\ O_2 \rightarrow 2.7\ CO_2 + 0.7\ H_2O + 0.2$ Soot + 34.1 MJ/kg intermediate

Soot + $2.66\ O_2 \rightarrow 3.66\ CO_2$ + 32MJ/kg Soot

Mixed Gases:

$0.572\ C_2H_4 + 0.383\ C_2H_6 + 0.043\ H_2 + 0.982\ O_2 \rightarrow$

$0.53\ C_2H_2 + 0.34\ C_2H_3 + 1.1\ H_2O$ + 14.2 MJ/kg

$0.61\ C_2H_2 + 0.39\ C_2H_3 + 2.66\ O_2 \rightarrow$

$2.66\ CO_2 + 0.813\ H_2O + 0.181$ Soot + 34.4 MJ/kg

Soot + $2.66\ O_2 \rightarrow 3.66\ CO_2$ + 32 MJ/kg

Technology for Reduction of Atmospheric Emissions in Petrochemical and Refining Industries 2014 Spring AIChE Meeting

SAS

Single-burner Mesh

Slide 17

- ➢ Rectangular cells
- ➢ Local refinement near burner tip
- ➢ 110,000 computational cells

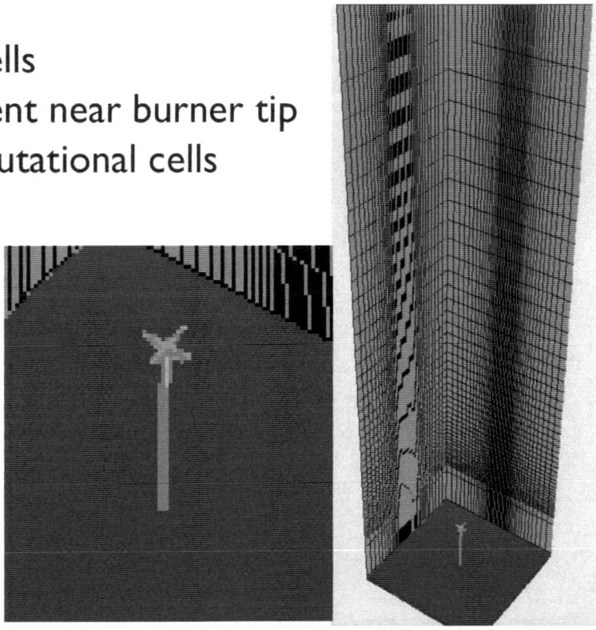

SAS

Single Tip Burning Propane: 3.0 m/s Wind

Slide 18

Front View

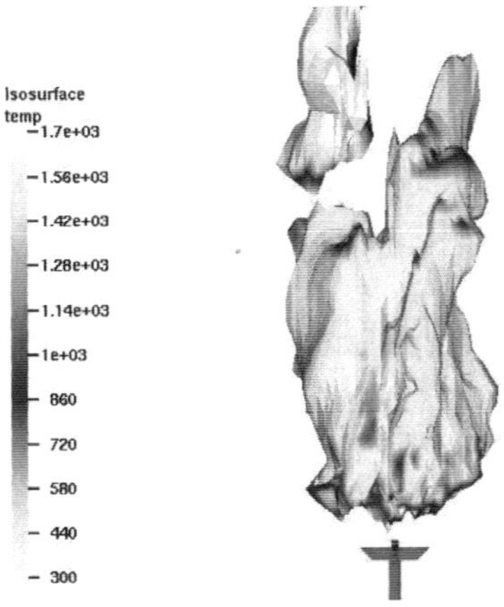

Side View

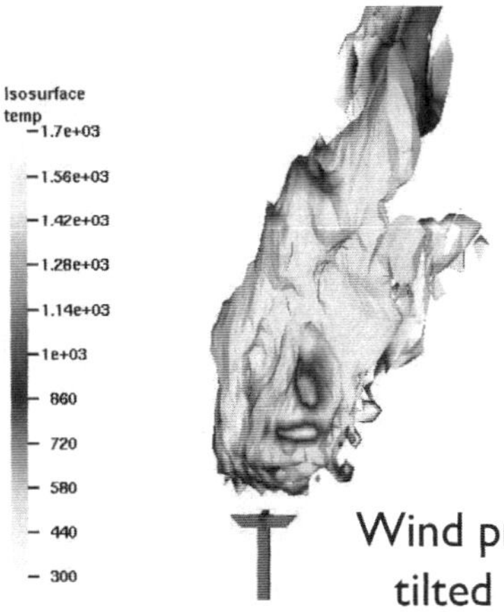

Wind produces tilted bushy, shortened flame

SAS

Single Tip Burning Propane: wind vs. no-wind

Slide 19

Single Burner

No wind produces tight "pencil-like" flames

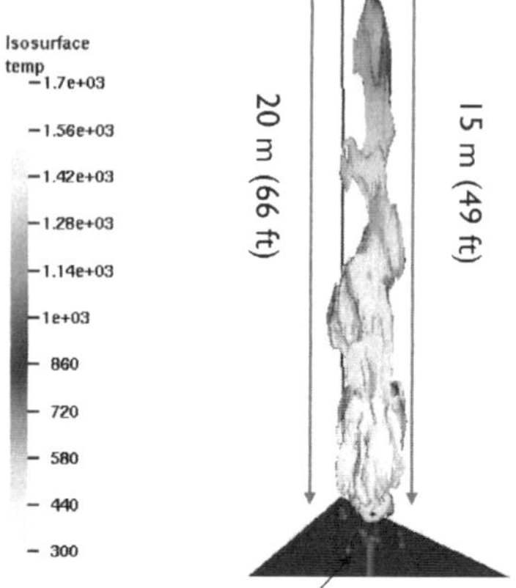

Flare tip located 2 m (~7 ft) above grade

Technology for Reduction of Atmospheric Emissions in Petrochemical and Refining Industries 2014 Spring AIChE Meeting

SAS

Single-Tip Ground Flare Test Results

Slide 20

- 1.4" WC @ 57 °F => 7.3 psig tip pressure
- 12-16 Km/hr (8-10 mph) crosswind
- ~30% flame height reduction
- Minimum flame tilt (~8°)

Technology for Reduction of Atmospheric Emissions in Petrochemical and Refining Industries 2014 Spring AIChE Meeting

Three-burner Mesh

- Rectangular cells
- Domain size is 30 m X 35 m X 25 m
- Local refinement near burner tips and radiation meters
- 188,000 computational cells

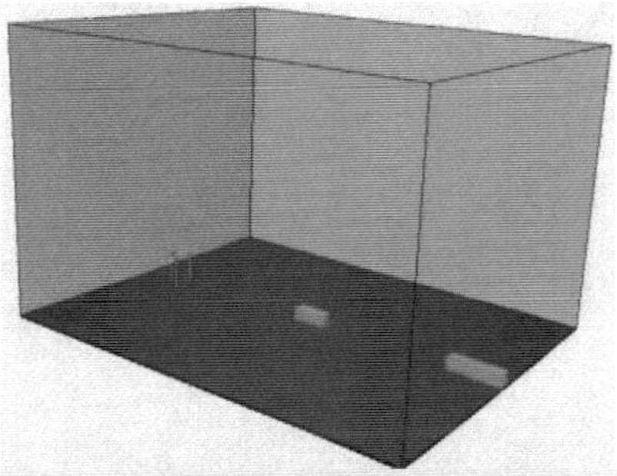

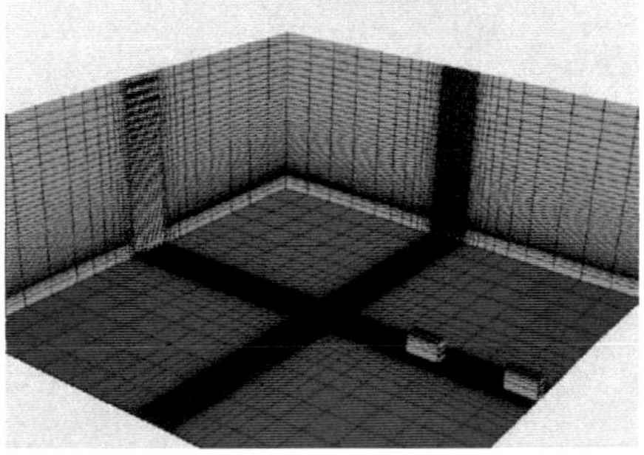

Predicted/Measured Flame for 3-burner Ethylene test

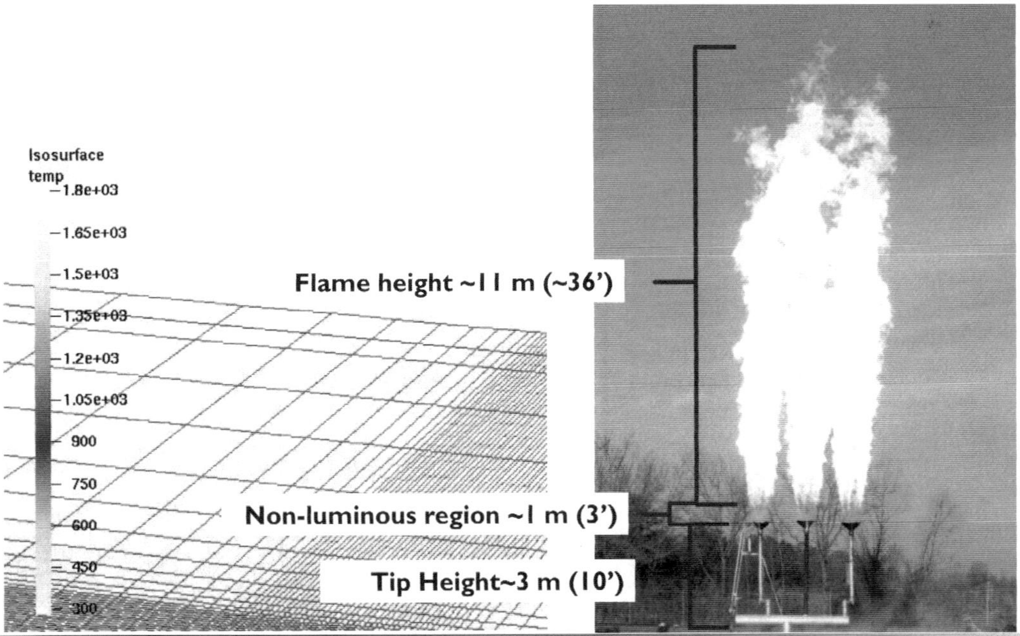

SAS

Predicted 3-burner flare with radiation monitors

ZEECO

Slide 23

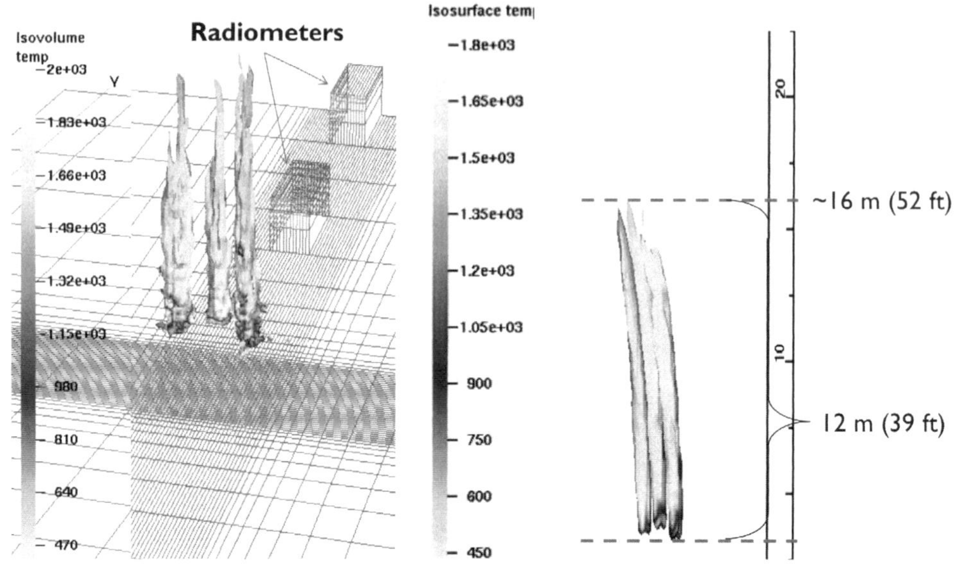

Smith, J.D., Suo-Ahttila, A., Smith, S., and Modi, J., "Estimation of the Air-Demand, Flame Height, and Radiation Load from Low-Profile Flare using ISIS-3D," *American – Japanese Flame Research Committees International Symposium, Advances in Combustion Technology: Improving the Environment and Energy Efficiency*, Marriott Waikoloa, Hawaii - Oct. 22 –24 (2007)

Technology for Reduction of Atmospheric Emissions in Petrochemical and Refining Industries 2014 Spring AIChE Meeting

SAS

Predicted Radiation from 3-Burner Flare with Modifications for Ground and Atmospheric Attenuation

ZEECO

Slide 24

Radiation Included in Prediction:

1. Ground re-radiating and reflecting incident radiation from flame
   - Assumed ground $\varepsilon = \alpha = 1$; allow ground to heat to steady state temperature
2. Atmospheric attenuation of radiation from flame
   - Model uses ambient/source temperatures with H_2O/ CO_2 absorption

WALL (plan view perspective)	Left Wall (W/m^2)	Right Wall (W/m^2)	Bottom Wall (W/m^2)	Flame Optical Thickness
PEAK FLOW				
Initial Radiation	78,000	63,000	108,000	0.275
Radiation Modification	61,000	35,000	NA	
SUSTAINED FLOW				
Initial Radiation	15,000	15,000	35,000	0.28
Radiation Modification	6,600	6,600	NA	

Technology for Reduction of Atmospheric Emissions in Petrochemical and Refining Industries 2014 Spring AIChE Meeting

SAS

Model Validation: Radiation Predictions Compared to Experimental Data

Slide 25

Tip Size	Position (m)	Burner Pressure (psi)	Total Predicted Radiation (W/m^2)	Measured Radiation (W/m^2)	Difference (%)
3	15	2.8	2700	3344	-20.0 %
3	15	7.3	4750	4803	-1.0 %
3	15	11.4	6150	6192	-0.7 %
3	30	2.8	650	671	-3.0 %
3	30	7.3	1350	1184	+14.0 %
3	30	11.4	1650	1532	+8.0 %
4	15	2.8	4325	6371	-32.0 %
4	15	7.3	8050	8192	-2.0 %
4	15	11.4	10000	9536	+5.0 %
4	30	2.8	1150	1513	-23.0 %
4	30	7.3	2580	2464	+5.0 %
4	30	11.4	3250	2747	+18.0 %

Smith, J.D., Suo-Ahttila, A., Smith, S., and Modi, J., "Estimation of the Air-Demand, Flame Height, and Radiation Load from Low-Profile Flare using ISIS-3D," *American – Japanese Flame Research Committees International Symposium, Advances in Combustion Technology: Improving the Environment and Energy Efficiency,* Marriott Waikoloa, Hawaii - Oct. 22 –24 (2007)

Technology for Reduction of Atmospheric Emissions in Petrochemical and Refining Industries 2014 Spring AIChE Meeting

SAS

Wind Effect on Radiative Flux

Slide 26

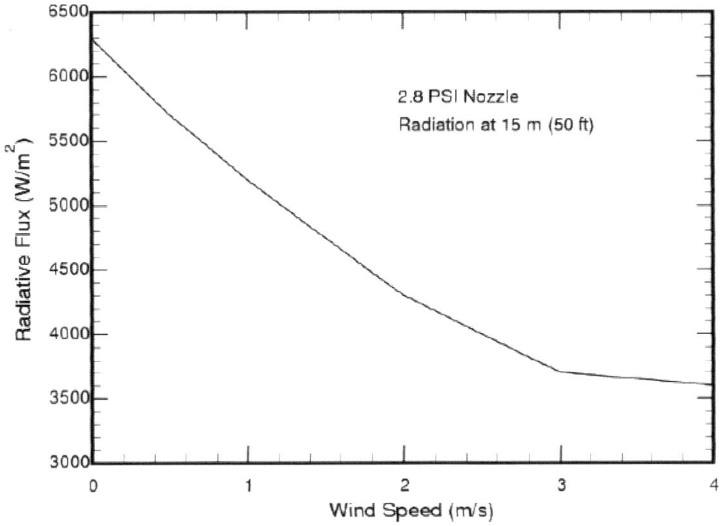

Smith, J.D., Suo-Ahttila, A., Smith, S., and Modi, J., "Estimation of the Air-Demand, Flame Height, and Radiation Load from Low-Profile Flare using ISIS-3D," American – Japanese Flame Research Committees International Symposium, Advances in Combustion Technology: Improving the Environment and Energy Efficiency, Marriott Waikoloa, Hawaii - Oct. 22 –24 (2007)

Technology for Reduction of Atmospheric Emissions in Petrochemical and Refining Industries 2014 Spring AIChE Meeting

SAS

Multipoint Ground Flare

Slide 27

Technology for Reduction of Atmospheric Emissions in Petrochemical and Refining Industries 2014 Spring AIChE Meeting

SAS

Full Field Flare Grid

Slide 28

- Domain size 10 m beyond wind fence; 25 m high
- Local refinement near burner rows/tips
- Sustained Flow Rate: 700,000 kg/hr
- Peak Flow Rate: 1,200,000 kg/hr

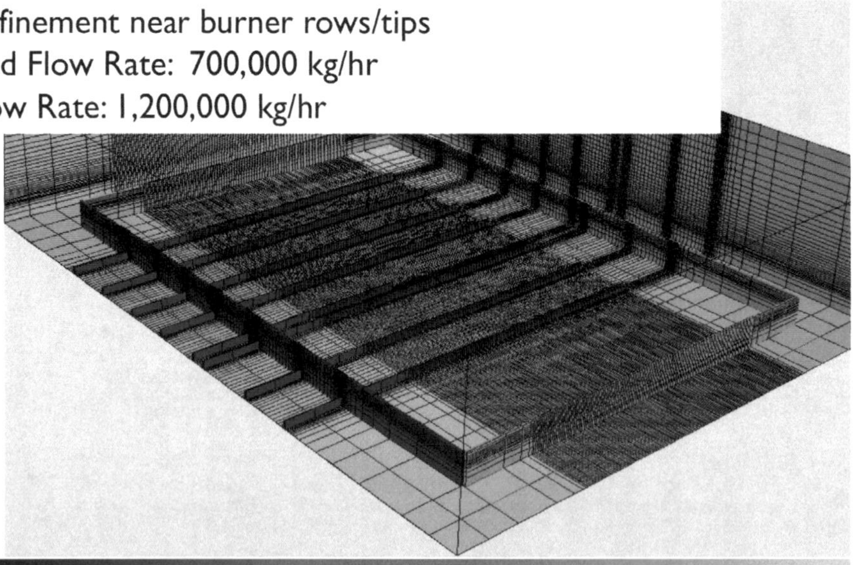

Technology for Reduction of Atmospheric Emissions in Petrochemical and Refining Industries 2014 Spring AIChE Meeting

SAS INCORPORATED

Simulation of Flare Tests

Slide 29

Predicted flame height well below total fence height

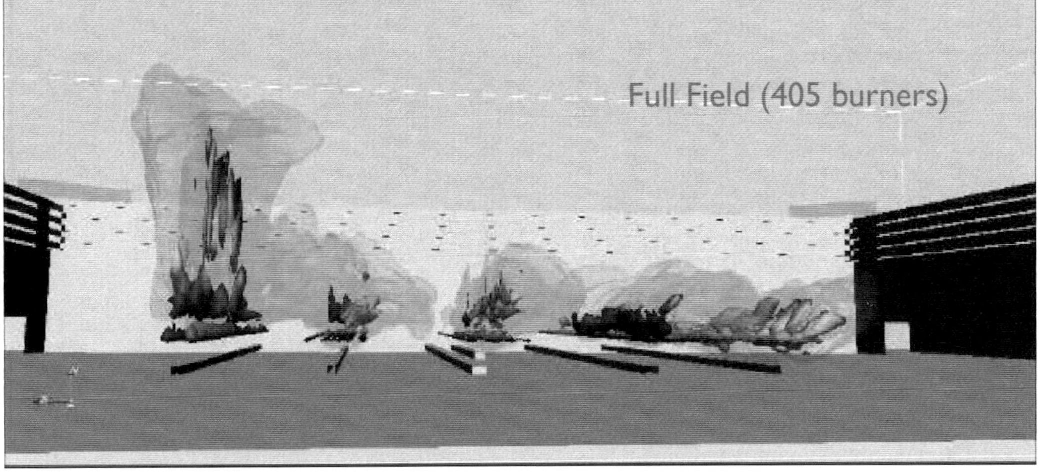

SAS INCORPORATED

Max $[C_2H_4]$ along line of sight for peak flow case

Slide 30

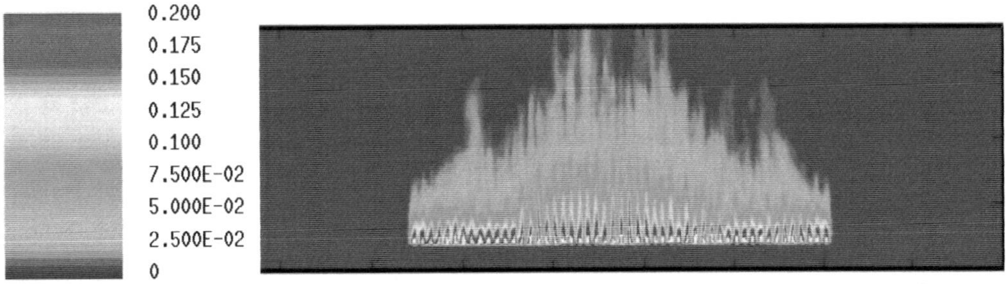

Side view along row showing flame elongation toward center of row

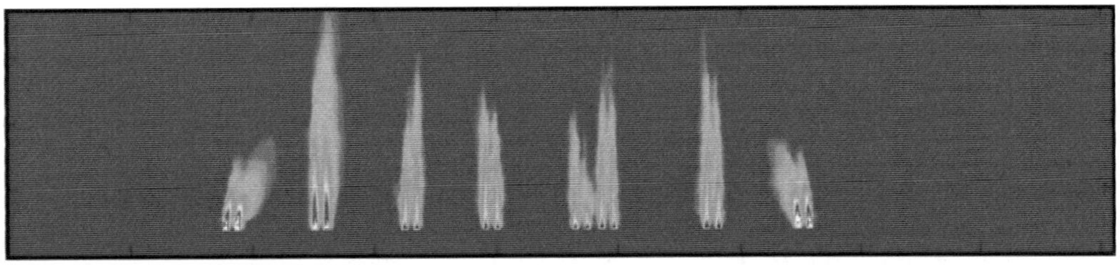

End view of flame height for each row and impact of inflow on outer rows

SAS INCORPORATED

Predicted Air/Fuel Requirements for "no-wind" cases

ZEECO

Slide 31

Case Description	Evaluation Plane Area (m^2)	Total Air Flow (kg/s)	Fuel Flow (kg/s)	Air/Fuel Mass Ratio
1 Burner Propane	14.63	60	0.46	130
1 Burner (Tip 3) Ethylene	13.26	52	0.94	55
3 Burner (Tip 3 - 7.3PSI) Ethylene	36	134	2.88	47
3 Burner (Tip 3 - 11.4PSI) Ethylene	36	144	3.84	38
3 Burner (Tip 4 - 7.3PSI) Ethylene	36	150	4.26	35
3 Burner (Tip 4 - 11.4PSI) Ethylene	36	160	5.79	28
Full Field Peak Flow Ethylene	3226	9700	262.3	37
Full Field Sustained Flow Mixed Gas	1843	4800	93.6	51.3

Technology for Reduction of Atmospheric Emissions in Petrochemical and Refining Industries 2014 Spring AIChE Meeting

SAS INCORPORATED

Conclusions

ZEECO

Slide 32

- CFD Flare Models:
  - Single-burner model: 110,000 cells
  - Three-burner model: 188,000 cells
  - Full-field model: 700,000 cells (Sustained Flow)/1,200,000 cell (Peak Flow) (Combustion chemistry for propane, ethylene, mixed gas considered)
- Analyzed flame shape/size for 3 fuels for single and three burner tests
- Compared predictions to measurements for 12 tests (2 tip sizes, 3 operating pressures, 2 radiation sample locations)
- "*Reasonable*" estimates of radiation heat transfer and air demand for low profile flare
  - Air/fuel ratios range from 28 to 47 for 3-burner test and from 37 (Peak Flow Case) to 51 (Sustained Flow Case)
- Calibrated flare model applied to full-flare field to estimate:
  - Air demand for specified tip/row spacing
  - Radiation load on wind fence for nominal and peak flow cases
  - Expected flame height and smoke production for nominal and peak flow cases

Technology for Reduction of Atmospheric Emissions in Petrochemical and Refining Industries 2014 Spring AIChE Meeting

SAS

Questions?

Slide 33

353103 A Multi-Period Process Integration to the Management of Process Flares with Consideration of Fuel Substitution and Cogeneration Systems

Abstract:

A wide variety of process upsets can lead to flaring. Such flaring activities cause environmental emissions and result in economic losses. A possible approach to mitigating these emissions and losses is to utilize the flared gases in combustion systems (boilers and industrial furnaces) that can be used to produce utilities such as steam. Additionally, the generated steam can be further used in a cogeneration system (e.g., through steam turbines) to produce power and heating. This paper introduces a process integration approach to the effective utilization of flared gases into combustion and cogeneration systems. Since the flared gases differ in composition and characteristics from the fuel that is typically used in boilers, there is a need to make sure that the fuel quality is in the acceptable range for the boiler. Various constraints may be used to assess usability of flared gases as part of the fuel blend fed to the boiler. For instance, the Wobbe index (WI) may be used to study fuel properties sent to the boilers. A multi-period optimization formulation is developed to determine the optimal usage of flare gases in the boiler and cogeneration systems. An operational period is discretized into multiple periods corresponding to different operating modes of the process (process episodes). Economic objectives are used while including technical and environmental constraints. Additionally, the cost-benefit analysis for adding a cogeneration system with an optimal size is carried out. A case study using an ethylene plant is used to demonstrate the various aspects of the process.

PROOF OF CONCEPT TEST FOR A REAL-TIME FLARE COMBUSTION EFFICIENCY MONITOR

Presented at
2014 AIChE Spring Meeting &
10th Global Congress on Process Safety

By

Yousheng Zeng, PhD, PE and Jon Morris
Providence Photonics, LLC

Mark Dombrowski
Surface Optics Corporation

April 2, 2014

PHOTONICS PROVIDENCE

A BRIEF REVIEW OF THE CONCEPT PRESENTED AT 2012 FALL CONFERENCE

Ref.: Zeng and Morris, "A New Method to Measure Flare Combustion Efficiency in Real-Time", presented at AWMA Louisiana Section 2012 Fall Conference, Baton Rouge, Louisiana, October 30-31, 2012

2

PHOTONICS PROVIDENCE

Flare Combustion Efficiency (CE):

$$CE(\%) = \frac{[C]_{CO2}}{\sum_i n_i[C]_{HCi} + [C]_{CO2} + [C]_{CO}} \quad \text{Eq. (1)}$$

Flare CE – Very difficult to measure

Source: TCEQ/UT

The Concept

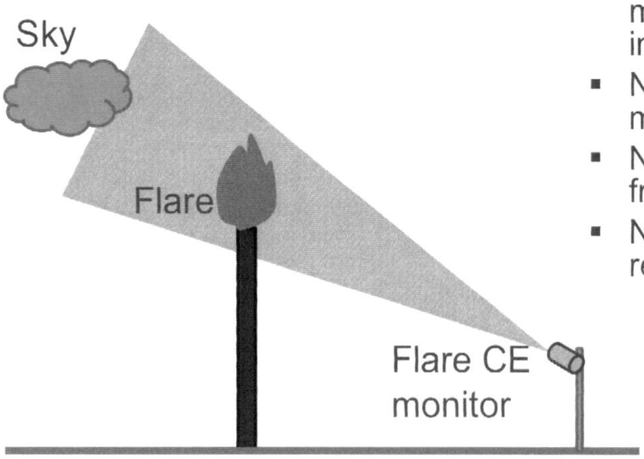

- Monitor and map flare CE in real time through a special multi-spectral IR imaging device
- Not a path measurement
- No scanning; high frame rate
- No operator required

We call it "Flare Efficiency Monitoring System" or "FEMS"

VISION

- Monitor flare CE in real-time
- Image the full flare flame; measure both overall CE and CE at a pixel level (CE mapping)
- Industrial grade device, suitable for integration with PLC or DCS
- One monitor covers multiple flares on site (step and stare)

PHOTONICS PROVIDENCE

WORKING PRINCIPLE

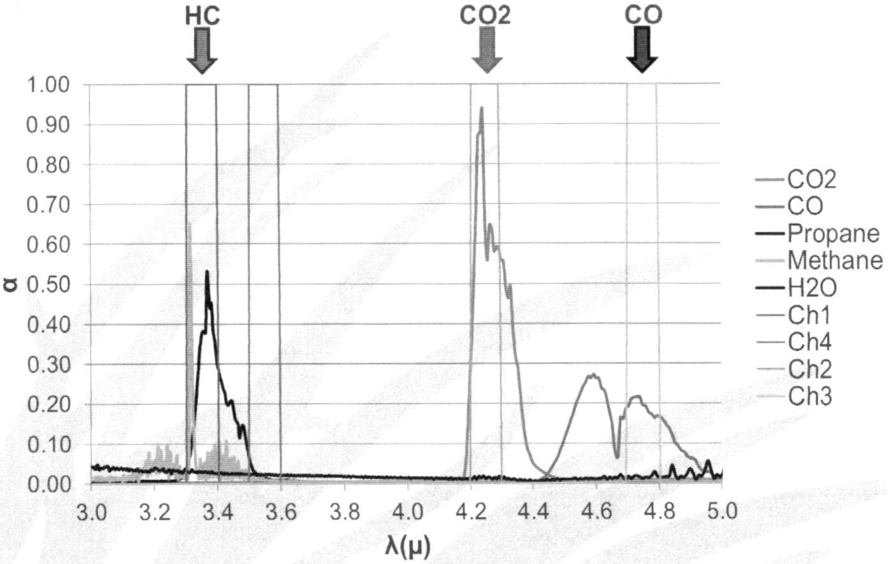

PHOTONICS PROVIDENCE

PROOF OF CONCEPT TEST

The proof of concept test was partially funded by EPA SBIR Phase I grant.

7

PROVIDENCE PHOTONICS

TEST SETUP

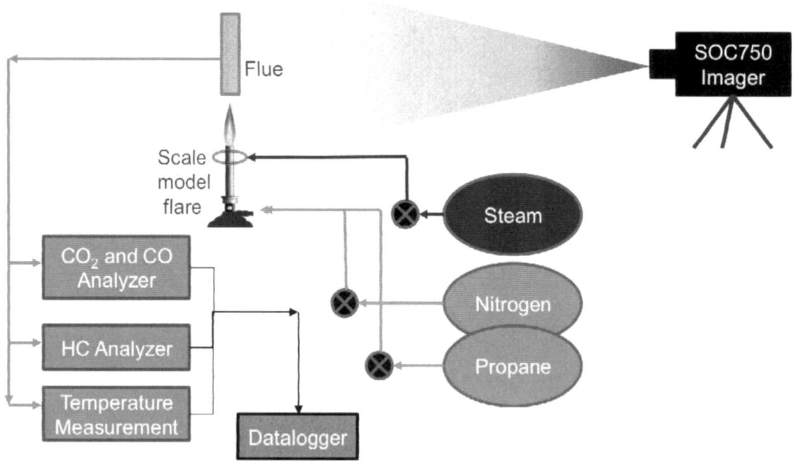

8

PROVIDENCE PHOTONICS

Distance from
the scale model flare
to the Imager: 23 ft.

SOC750
Hyper-Spectral
Imager

Scale model flare,
Flue w/ Sampling Probe, and
Analyzers

SCALE MODEL FLARE

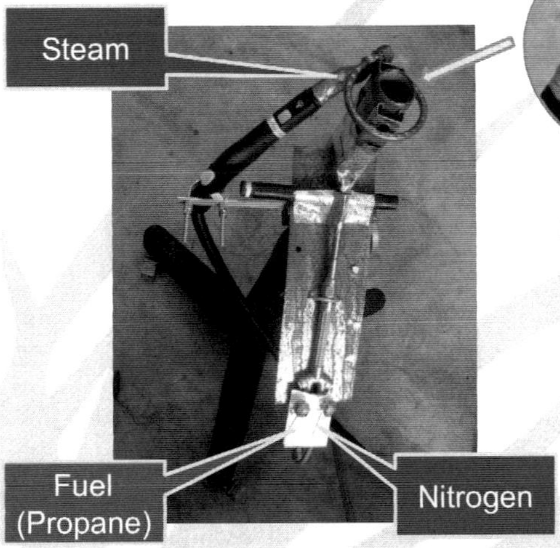

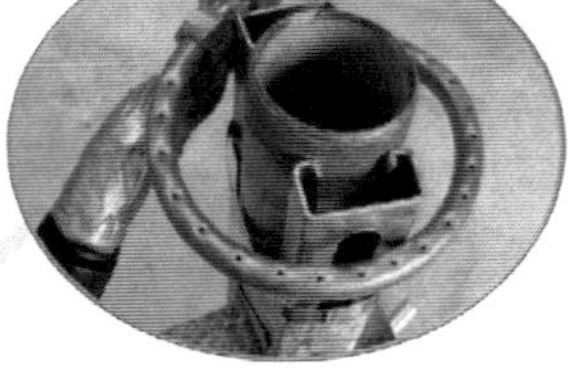

Fuel: approx. 40k Btu/hr.
Nitrogen: varied
Steam: ~ 3 lb./hr.

FLUE AND SAMPLING PROBE

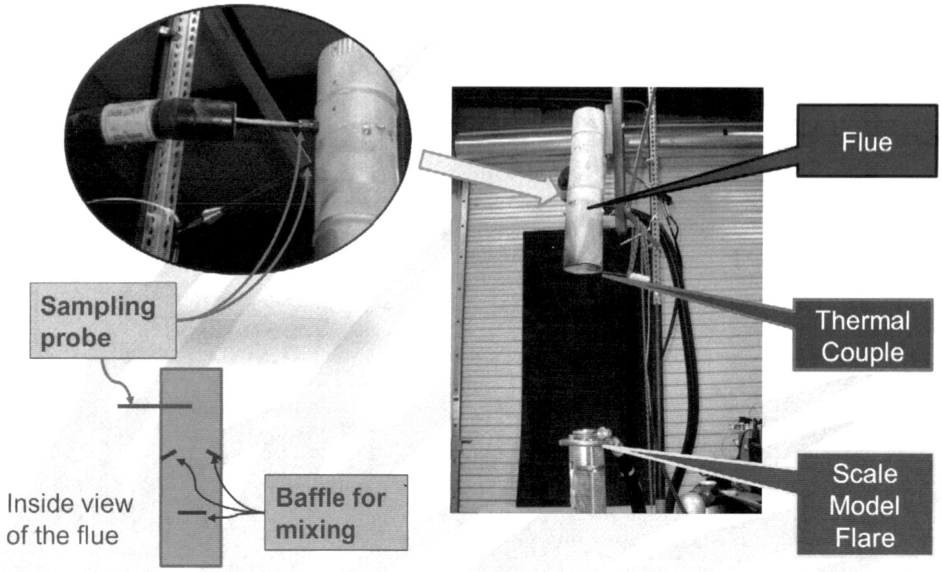

ANALYZERS AND IMAGER

Testo 350 XL
(CO2 using NDIR, CO, O2, NOx, H2, Temp, etc.)

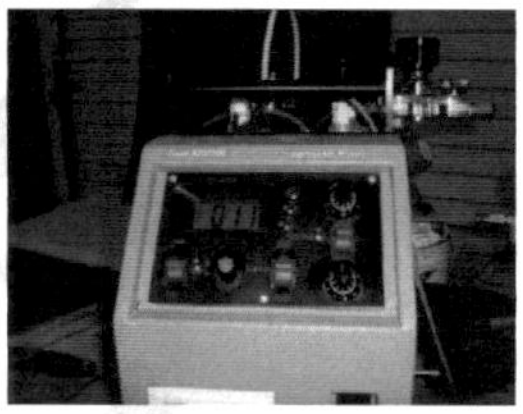

3010 Mini FID
Calibrated to Propane

SOC750 Hyper-spectral imager
42 spectral channels, operated at 22 cubes per sec.

TEST RUNS

PHOTONICS
PROVIDENCE

PRELIMINARY RESULTS - WITHOUT CALIBRATION

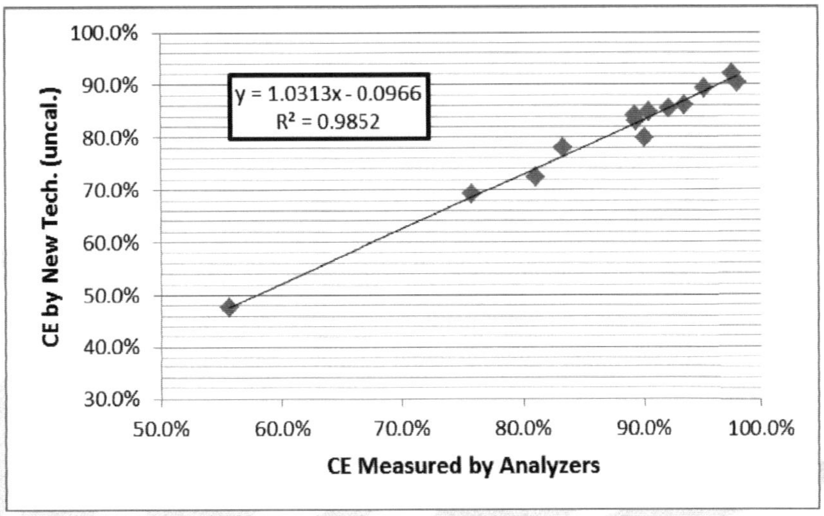

PHOTONICS
PROVIDENCE

PRELIMINARY RESULTS - WITH AN INITIAL CALIBRATION

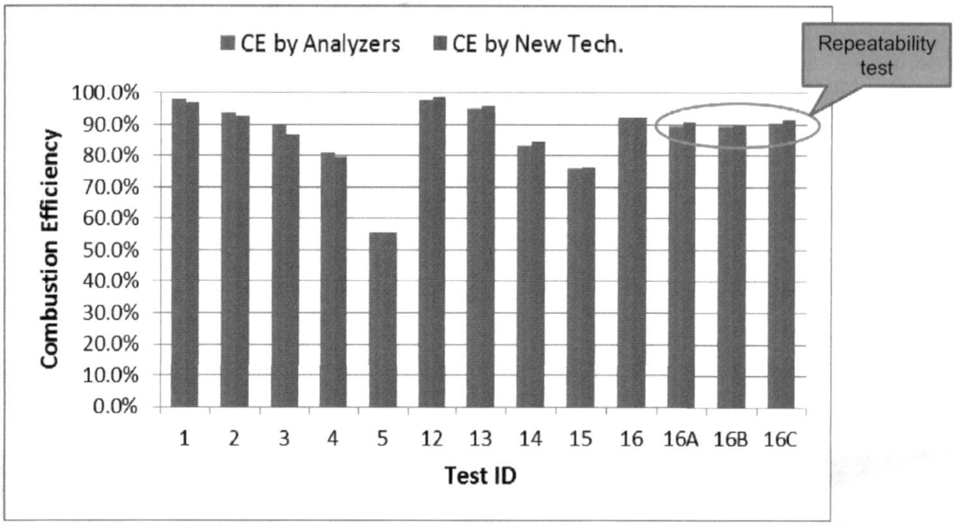

15

PHOTONICS PROVIDENCE

PRELIMINARY RESULTS - INITIAL METHOD TO MEASURE TEMPERATURE

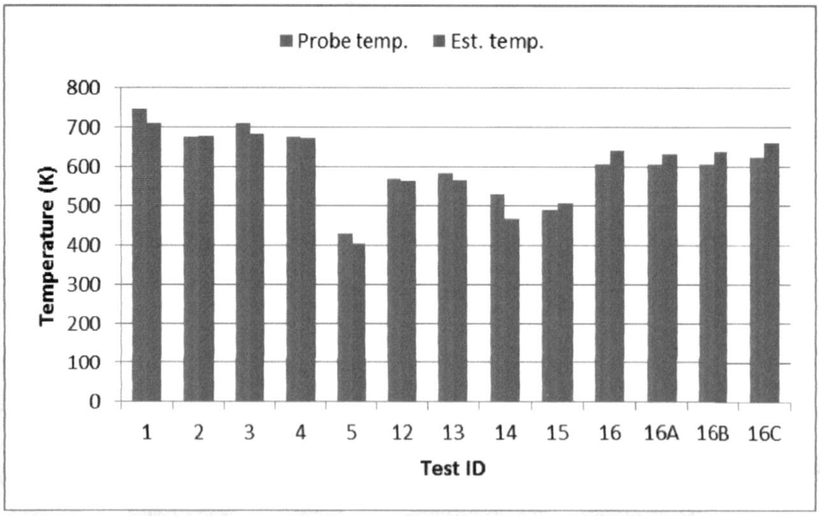

16

PHOTONICS PROVIDENCE

PRELIMINARY RESULTS – CE MAP

30 sec. average (670 frames)

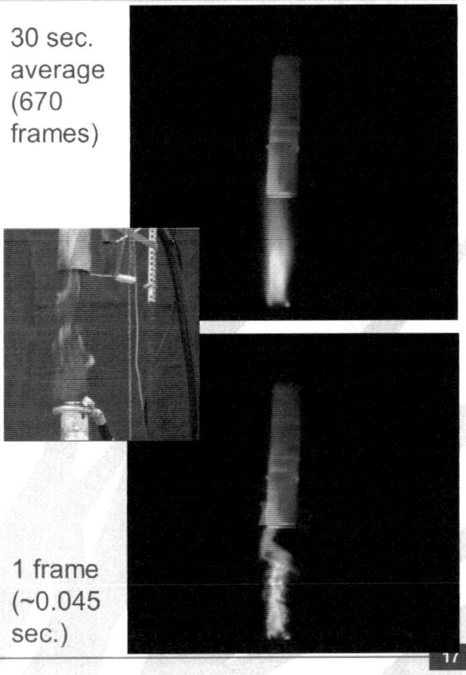

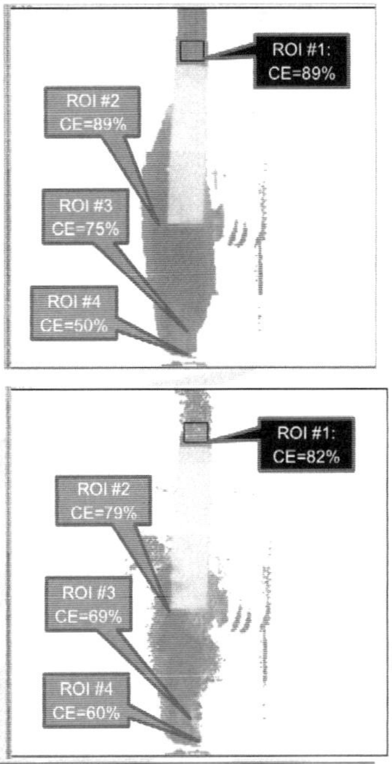

1 frame (~0.045 sec.)

Three Types of Measurement

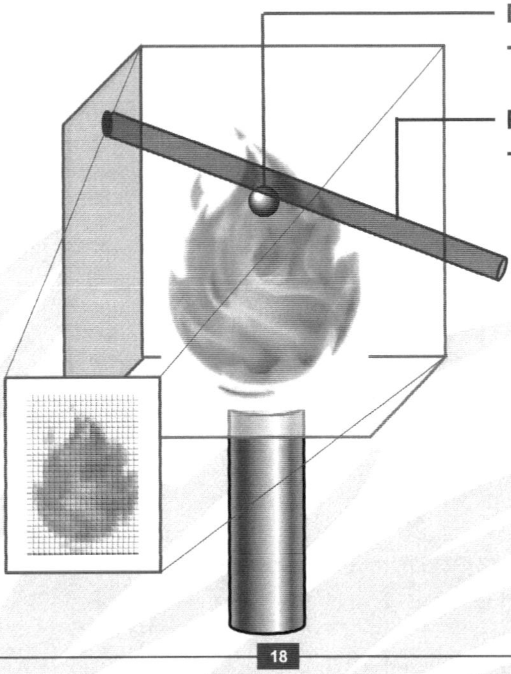

Comparison with PFTIR

PFTIR

- "Scanning" - >1 sec/scan – assuming that flare is static during that time
- Path measurement – aiming required
- Human operation

FEMS

- Staring – 20-30 data cubes/sec – match the flare dynamics
- 2-D mapping of CE – no aiming required
- Automatic

19

PHOTONICS PROVIDENCE

Conclusion

- The CE determined by the new technology correlate well with the CE measured by conventional analyzers
- With a further developed calibration method, real-time CE monitoring and feedback for flare optimization is feasible
- The new technology can determine CE at a pixel level, generating a CE map for the entire flare flame. No aiming issue.
- As a side benefit, it can also provide temperature mapping of the flare flame

20

PHOTONICS PROVIDENCE

LOOKING FORWARD…

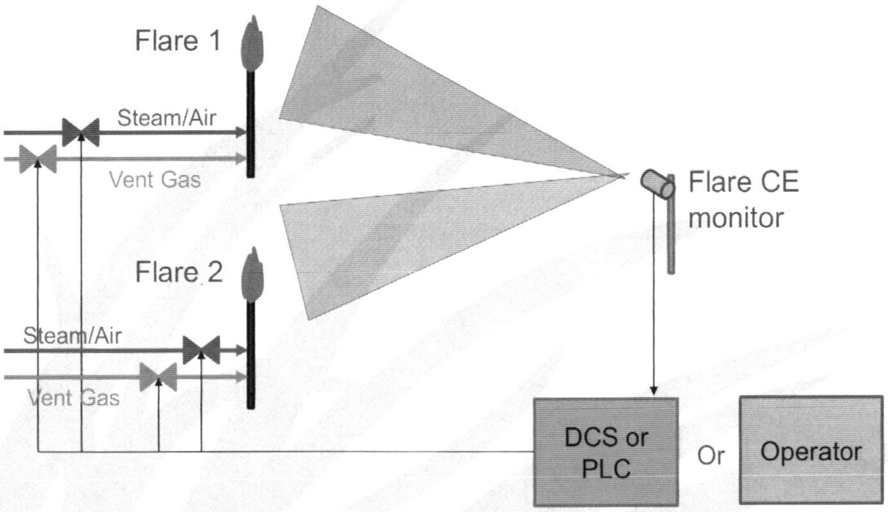

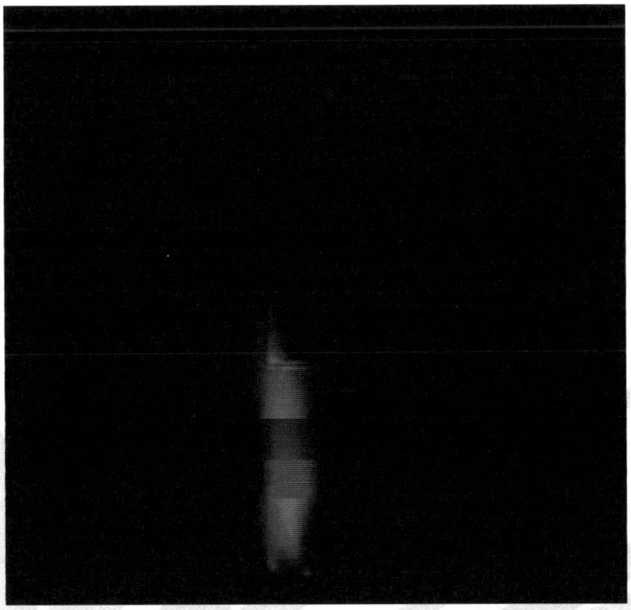

Contact Info. youshengzeng@providenceeng.com

PHOTONICS PROVIDENCE

348171 Flameless Thermal Oxidation

Abstract:

Linde's patented Flameless Thermal Oxidizers (FTO) technology oxidizes combustible gases and atomized liquids with air under controlled conditions in a specially designed reactor without a flame. Residence time and reaction temperatures are carefully selected to achieve 99.99 +% destruction efficiency with ultra-low NOx and dioxin emissions at very high thermal energy efficiency. These systems are commercially proven, cost effective and highly effective for the destruction of volatile organic compounds (VOC's) and hazardous air pollutants (HAP's). This paper will present an overview of the FTO technology and market applications with special emphasis on environmental performance and cost considerations.

350916 Modeling of the Dynamic Interaction in an LDPE Closed System Purge Bins-Regenerative Thermal Oxidizer (RTO) for Ethylene Emission Mitigation

Abstract:

Common designs of LDPE purge bins involve venting the purge and blend air to atmosphere containing small traces of ethylene. To mitigate ethylene emission from these bins, a new process design has been proposed to collect vent air-carrying ethylene via a closed ducting system connected to a regenerative thermal oxidizer (RTO) unit. This necessitates a blower at the downstream end to pull the purge air before feeding into the RTO to incinerate the ethylene. Concerns have been raised regarding the impact of such system on the purge bins, which are not designed to withstand any significant amount of vacuum. While conceptually steady operation of such system is feasible, operation of purge bins is never steady particularly when the purge bins are not designed to withstand any significant level of vacuum pressure. With the minimum safe operating pressure at -0.1 psig (-0.68 kPag), and the design RTO inlet pressure being -1"'WC (-0.25 kPag), the bins are forced to operate under slight vacuum conditions with a very narrow allowable margin for error. The operation is inherently dynamic as LDPE pellets are being conveyed into and out of the bins, purged and mixed in cycle by the interaction of multiple pieces of equipment. In the event of any equipment failure along the train, pressure in the bins can drop below their minimum vacuum level and cause damage to the shell. Dynamic interactions between the various components of the system including the blower with its VFD drive, purge and blend air compressors, capacitances and resistive characteristics of the pellets in the bins, as well as the response of the control system all contribute to interesting dynamic interactions that must be attended to. A model capable of simulating various dynamic phenomena has been built using Aspen HYSYS Dynamics TM v7.3 for the entire system, and used to answer what-if operational scenarios. As an example it was shown that the addition of a pull blower with its variable frequency drive (VFD), cannot maintain the bin pressure above the vacuum pressure limit of -0.1 psig even with the installation of individual pressure vacuum relief valve (PVRV) on each bin. Any interruption to air flows such as the shut-down of blend blower(s), stop/start conveying or swinging the purge air blower poses a non-negligible risk. Interaction between the PVRV's and the blower's VFD both simultaneously trying to control the bin pressure could lead to pressure fluctuations that exceed the design vacuum pressure of the bins. This paper will provide an overview of the system, state the design and operational challenges, basis of the dynamic model, and results of several what-if scenarios.

351659 Air Quality Impact of the Startup of a Single Olefin Plant

Abstract:

Ground-level ozone is one of six common pollutants defined by US EPA (Environmental Protection Agency). It can cause health effects, including lung diseases. HRVOC (Highly Reactive Volatile Organic Compounds) emissions from industries are highly responsible for ground-level ozone pollution, especially around industrial district. Flare emissions from MSS (maintenance, startup and shutdown) & events are major sources of HRVOC emissions.

Most of air quality modeling treats industrial flares as constant emissions, which means the emissions keep constant for the whole scenario days, and the total amount of emissions equal to the total emissions reported to environmental authorities. However, real flare emissions from MSS and events are time-various. At certain time, flare emissions could be very huge, and at other time, flare emissions could be very small. During emission peaks, the ozone formation could be much higher than a flat emission curve.

In this paper, time-various flare emissions of startup procedures are considered for air quality modeling. Emission data are obtained from dynamic simulation of a typical ethylene plant. The dynamic simulation results were validated with real plant data and emission data of startup events. The time-various flare emission data were added into Houston-Galveston-Brazoria (HGB) 2006 episode base case modeling as a point emission source. Three startup scenarios were simulated, which include a baseline scenario and two flare reduction scenarios. For each scenario, 120 cases were built to simulate different starting time covering the whole episode days.

Simulation results show that for the baseline scenario, the maximum 8-hour average ozone increment could be 15 ppb at certain starting time. It means starting a single plant could temporary increase the regional ozone concentration by 15 ppb, which is 20% of current ozone standard (75 ppb). For the two flare reduction scenarios, the maximum 8-hour average ozone increment could also reach 10 ppb. Simulation results also show that ozone increment is high when starting time is before 8 am. Ozone increment is low when starting time is after 12 pm. For each scenario, air quality simulation gives least impact period, which means ozone impact could be low when starting time is within that period. For the baseline scenario, the least impact period is only 2 hours. For the two flare reduction scenarios, the least impact periods are 4 hours and 8 hours.

In conclusion, dynamic simulation and air quality simulation show that the flare emissions from MSS and events are highly time-various, and may not be treated as flat emissions. Air quality simulations also give least impact periods of startup time, which could help chemical plants to arrange their startup schedule to reduce ozone formation.

SOLID FUEL TO NATURAL GAS CONVERSIONS FOR EXISTING BOILER APPLICATIONS

William Gurski
Director Power Sales
Zeeco Inc.

John Guarco
Technical Director Boiler Burners
Zeeco Inc.

Nando Nunziante
Manager Power Controls
Zeeco Inc.

Prepared for Presentation at the Year 2014 Spring National Meeting
New Orleans, LA, March 30 – April 3, 2014

Abstract

Recent discoveries of vast natural gas reserves in the United States have resulted in increased production of Natural Gas, resulting in lower prices. Petrochemical facilities are performing solid fuel conversions on their boilers to natural gas as a cost-effective and efficient fuel solution. Natural Gas is not only economically beneficial but also environmentally efficient with cheaper prices and reduced SO_2, NO_x, and CO_2 emissions. As more facilities consider boiler fuel conversions, it is important to understand all facets of the conversion, from the thermal evaluation of the boiler, to the complete design, supply and installation of the new firing system.

We will provide specific details and recommended practices from a recent Circulating Fluidized Bed (CFB) Boiler fuel conversion application designed for 1.4 billion Btu/hr of heat input for the maximum continuous steam rating. We will also review thermal performance results, design and supply of the complete new gas firing system, and installation conversion assistance for the boiler modifications and firing system installation details.

Introduction

Zeeco was contacted by a large industrial plant owner to assist them with the fuel changeover of a utility designed solid fuel CFB. Due to current contract requirements and the competitive nature of the industry, the plant owner will remain confidential.

The primary goals of the project are as follows:

- Full Conversion from Petcoke firing to Natural Gas
- Maintain boiler steam design capacity (~1200 kpph)
- Meet permitted NO_x requirements – 0.07 lbs/MMBtu
    - 25-100% heat input for NO_x emissions
- Minimal or no impact on the boiler waterside design

The owner provided the original boiler design information for Zeeco's review of potential boiler burner retrofitting options. This was the first of two boilers that would require retrofitting Two sister units were built/commissioned on solid fuels approximately three years previously. The units were originally designed with natural gas start-up burners and commissioned for main fuel support via the petcoke. The petcoke was to be fired and combusted via a circulating fluid bed system. Upon initial advisement from the plant owner, they noted that their petcoke fuel contract was no longer cost effective, as this fuel feedstock was significantly higher in price than originally estimated in the plant system's budget. The boilers were only in operation for less than two years when this decision to switch to natural gas was determined.

Zeeco will review the major facets of this boiler conversion—from the initial proposal phase to the commissioning and startup of the boiler and firing system. We will also discuss the combustion parameters of the technology behind the project's success–Zeeco's DT Ultra-Low NO_x Free-Jet Burner technology.

Combustion Parameters

In order to understand why the ZEECO® Free-Jet technology was chosen, formation of thermal NO_x emissions must first be examined. For gaseous fuels with no fuel-bound nitrogen (N_2), thermal NO_x is the primary contributor to overall NO_x production. Thermal NO_x is produced when flame temperatures reach a high enough level to "break" the covalent N_2 bond apart, allowing the "free" nitrogen atoms to bond with oxygen to form NO_x.

Stoichiometric Equation describing typical combustion in a natural gas fired burner:

Methane & Air with Excess Air

$$2CH_4 + 4\ (XA)\ O_2 + 15\ (XR)\ N_2 \dashrightarrow 2CO_2 + 4H_2O + (XA)\ 15N_2 + (XR)\ O_2$$

Natural air is comprised of 21% O_2 and 79% N_2. Combustion occurs when O_2 reacts and combines with fuel (typically hydrocarbon). However, the temperature of combustion is not normally high enough to break all of the N_2 bonds, so a majority of nitrogen in the air stream passes through the combustion process and remains diatomic nitrogen (N_2) in the inert combustion products. Very little N_2 is able to reach high enough temperatures in the high intensity regions of the flame to break apart and form "free" nitrogen. Once the covalent nitrogen bond is broken, the "free" nitrogen is available to bond with other atoms. Basic chemistry dictates that free nitrogen, or nitrogen radicals will react to other atoms or molecules that can accept them to create a more stable atom. Of the possible reactions with the products of combustion, free nitrogen will most likely bond with other free nitrogen to form N_2. However, if a free nitrogen atom is not available, the free nitrogen will react with the oxygen atoms to form thermal NO_x. As the flame temperature increases, the stability of the N_2 covalent bond decreases, allowing the formation of free nitrogen and subsequently increasing thermal NO_x. Burner designers can reduce overall NO_x emissions by decreasing the peak flame temperature, which can reduce the formation of free nitrogen available to form thermal NO_x.

The varied requirements of Power and Industrial plants entail the use of numerous types and configurations of burners. The method utilized to lower NO_x emissions can differ by application. However, thermal NO_x reduction is generally achieved by delaying the rate of combustion. Since the combustion process is a reaction between oxygen and fuel, the objective of delayed combustion is to reduce the rate at which the fuel and oxygen mix together and burn. The faster the oxygen and the fuel gas mix, the faster the rate of combustion and the higher the peak flame temperature.

Figure 1 plots Peak Flame Temperature against Thermal NO_x created. NO_x emissions increase as the adiabatic flame temperature increases. Slowing the combustion reaction reduces the flame temperature, which results in lower thermal NO_x emissions. The challenge in achieving lower thermal NO_x emissions is not the theory; however, it is in retrofitting the latest burner technologies into older existing boilers without adding expensive external components or processes.

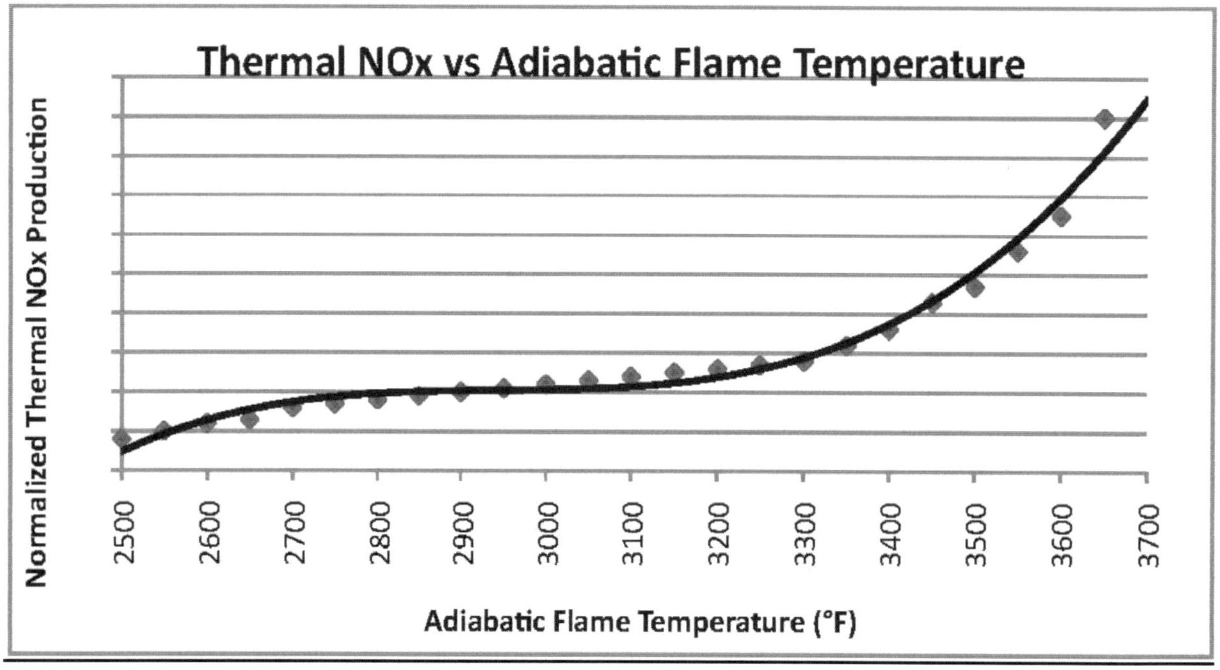

Figure 1: Calculated Peak Flame Temperature vs. Thermal NO_x Production

The industry's standard method to reduce thermal NO_x is to mix the fuel gas together with the inert products of combustion to recondition the fuel before combustion occurs. Since the reconditioned fuel is mainly comprised of inert components, the resulting composition burns at a lower peak temperature. To best utilize the inert products of combustion (flue gas) within the boiler, the fuel gas is introduced along the outside perimeter of the burner tile in an area where flue gas is present while the boiler is in operation. As the fuel gas passes through the inert products of combustion, mixing naturally occurs. This changes the composition of the fuel, and stabilization occurs at the tile exit. Since the reconditioned fuel mixture is 80 to 90% inert in most cases, the resulting flame burns at a lower peak temperature and generates less thermal NO_x.

The mixing of the fuel gas with flue gas prior to combustion is called Internal Flue Gas Recirculation (IFGR). When IFGR is too aggressive, it can result in an increased blower power usage, decreased burner turndown, and increased flame destabilization. Through Free-Jet Theory, maximizing IFGR while maintaining flame stability and flame length can become a challenge.

Flame Stabilization

In order to fully utilize the Free-Jet Theory, Zeeco devised a method to stabilize the burner flame with a highly inert fuel gas/flue gas mixture. This type of combustion is achieved when the flame is stabilized in a low-pressure area created on a series of specially designed hot refractory ledges. As combustion occurs, the refractory ledge retains heat and flame stability is enhanced. Thus, to achieve improved stability and extreme Thermal NO_x reduction, the Free-Jet Technology:

1. Mixes inert flue gas through free-jet methods with all of the fuel gas before combustion occurs, lowering flame temperature.
2. Stabilizes the flame on a refractory ledge, improving flame characteristics

Before combustion is initiated, a furnace is typically filled with normal air, which contains 21% oxygen. Once the burner is ignited, the oxygen content inside the furnace decreases until the burner achieves maximum duty. At this point, the oxygen content in the furnace is normally 2 to 3%. To keep the burner stable throughout the transition from start-up with 21% oxygen to maximum duty with 2 to 3% oxygen, Zeeco developed a series of stabilization ledges as shown in **Figure 2**. These ledges are a design feature of the Zeeco® Free-Jet Burner chosen for the boiler retrofit.

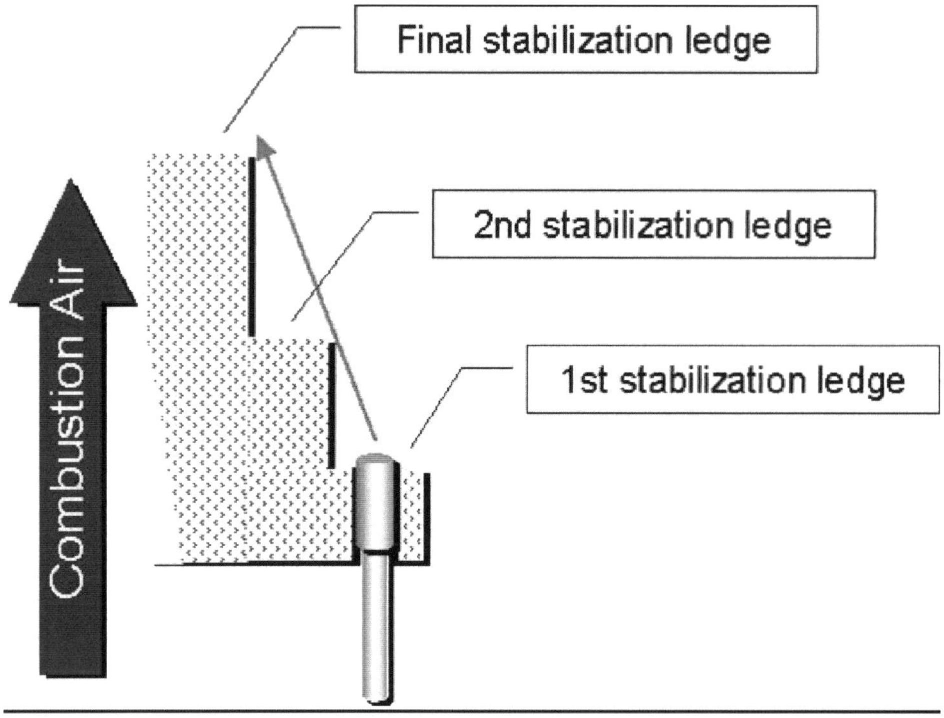

Figure 2: Free-Jet Flame Stabilization Method Illustration

Burner Technology Description

Introduction to Free-Jet Technology

The Zeeco Free-Jet burner series was designed with the specific purpose of maximizing the amount of IFGR to reduce thermal NO_x emissions without sacrificing burner performance with respect to flame length, turndown, and stability. The maximization of IFGR means many of the problems associated with using high levels of EFGR to achieve low emissions can be reduced or eliminated. Specifically, Zeeco's Free-Jet design dramatically reduces or eliminates the need for EFGR by reducing blower power usage, increasing turndown, reducing maintenance and improving flame quality.

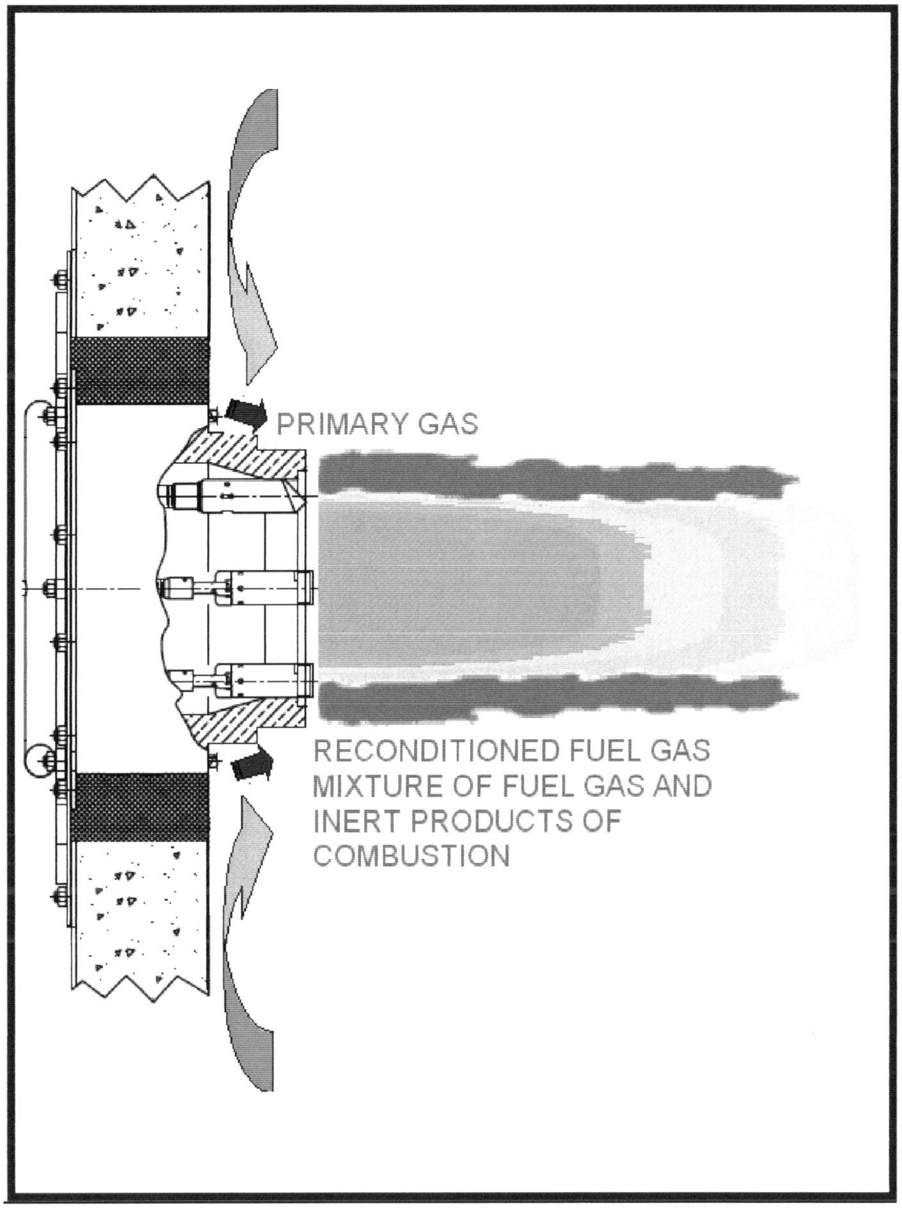

Figure 3: Inert Products of Combustion (Flue Gas)

Pre-Award Engineering

Since the boiler was only in operation for a short period of time, the plant needed assistance with the selection of the new gas firing system and options that would allow them to operate the boiler at the same steam flow capacity and pressure as originally designed for petcoke firing. Zeeco designed three possible boiler conversion options for the operator to consider. Every option required removing the existing start-up burners and closing off the tube openings or retrofitting with new larger burners:

1) Remove the fluid bed bottom and vertically fire the boiler. This option was feasible for a full 100% conversion as the entire fluid bed and nozzle/air system would be removed in favor of six new burners with a corresponding windbox and air system (1500 MMBtu/hr).
2) Removing the four start-up burners and increasing the firing capacity to the necessary levels for full load steam conditions (1500 MMBtu/hr).
3.) Removing the four start-up burners with new burners, plus adding a second level of four burners above to achieve the necessary heat input for full load steam conditions (1500 MMbtu/hr).

Option One was not selected, as it would have required an entire new lower section of the boiler, plus major reconstruction of the fluid bed in order return to solid fuel firing. Option Two was not selected, as the heat release/absorption area within the boiler would have been too great for the water/steam production side.

Option Three was selected, as the owner determined that they wanted the flexibility to return to petcoke firing should the price for natural gas rise to a level higher than petcoke fuel. This allows ~80% of the system's equipment to remain available for future petcoke firing. This solution also offered the least cost to retrofit back to solid fuel firing. The plant owner also had the challenge of the water/steam side of the project, as the boiler system would now be required to operate on 100% natural gas firing instead of a high pressure recirculating fluid bed system. Zeeco assisted the plant owner and the boiler's Original Equipment Manufacturer (OEM) with the system design and selection. Several iterations were submitted by the Boiler OEM on the redesign, and Zeeco worked as a team with the three parties to select the best option for firing the boiler.

The challenge with the boiler redesign interactions was to ensure that the following conditions were met, as required by the plant owner:

- Maintain main steam flow per original design
- NO_x emissions at permit levels
- Boiler able to follow steam needs per electrical generation requirements
- Ability to ramp per the original design conditions
- Turndown of 10-1
- No design impact or changes on existing water or steam circuit metallurgy (tube material changes)

Two design challenges were discovered throughout the Pre-Award Engineering Phase. The first included the air system redesign due to the fluid bed operating on primary and secondary air systems. This is further detailed in the Contract Execution Section below. Another challenge was the integration of the natural gas firing Burner Management System (BMS) and controls into the existing Distributed Control System (DCS). This is also further explained in the Contract Execution Section.

Project Engineering

Once awarded, Zeeco immediately began the burner and BMS systems design. The plant owner required less than 26-week delivery for all components, as the need to fire natural gas was critical to the financial status of the project.

Zeeco began the contract with a full-site visit with the entire team, including the plant owner and boiler OEM. The technical requirements and time schedule were reviewed to ensure all those involved were informed of the critical nature of the project. The technical details of the project varied, as the unit was being fully converted from a solid fuel CFB to a natural gas fired utility boiler. Many items required detailed analysis during the contract phase, including:

- Burner integration into the existing waterwall and boiler structure
- Addition of one new elevation of burners and effects of flame in this furnace region
- Complete redesign of the burner combustion air system and air heater outlet to a single pressure and delivery system to the new burners
- Integration of the new BMS panel and controls logic into the existing DCS system
- Design and location of all new gas fired fuel valving and rack systems, plus local termination cabinets
- Refractory removal and modification to support new natural gas firing (lower boiler refractory was no longer needed due to the elimination of petcoke and recirculating solids from the cyclone)
- Structural modifications to support all equipment and ancillaries for the natural gas firing (combustion air ducting, burner supports, valve racks, etc.)

Burner integration posed several challenges. The new burners were required to fit into the existing waterwall opening and modified from a downfired arrangement to a horizontal firing arrangement. Zeeco advised the plant owner that the new Zeeco Free-Jet DT burner would be able to fit within the existing opening, and would allow the exact tube opening design to be used for the new upper burner elevations. Zeeco also assisted with the design of the support systems and relocation of all structural steel for the new burner locations (maintenance platforms, etc.).

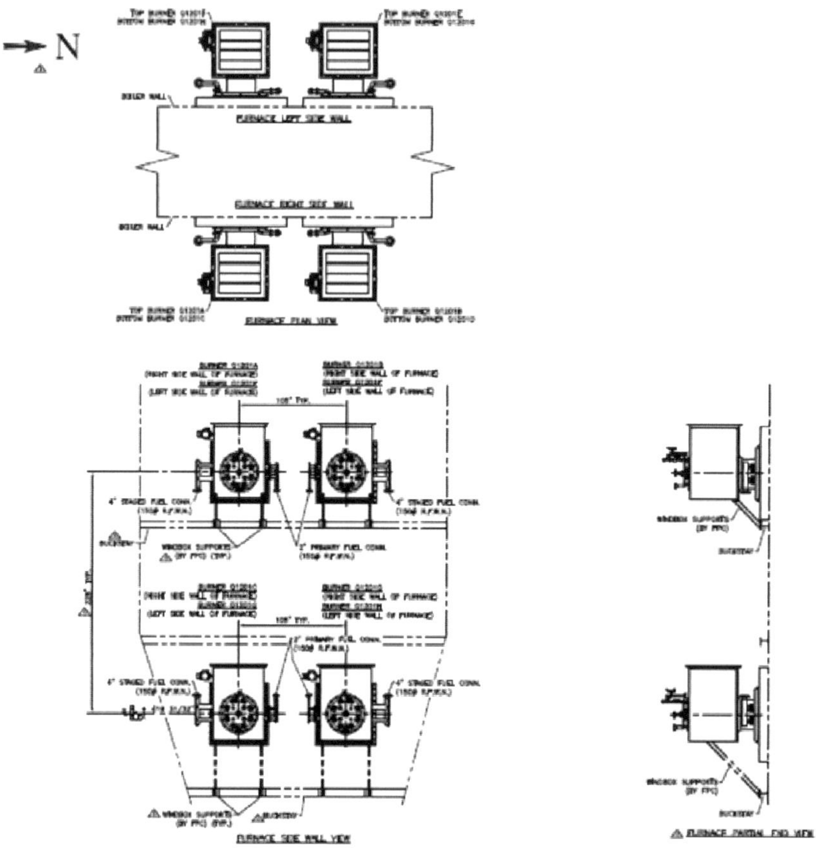

Figure 4: Boiler Design Diagram

It was important to ensure the new upper burner elevation had a proper flame fit within the boiler furnace and would not adversely affect heat transfer to the waterwalls or radiant furnace super heat. An elevation of ~19 feet was selected for the new burners above the existing burners as the ideal location, for not only proper flame fit but also ease the installation of the burners without major boiler structural modifications.

Figure 5: Zeeco's DT Free-Jet Burner

The combustion air system was completely redesigned from a two-system design (primary for the fluid bed nozzles and secondary air system for bed support), to a single pressure design. Zeeco was contracted to assist the plant owner with the entire system redesign, as well as routing to the new burner elevations. Zeeco also assisted with a new design for the air heater outlet and fan control due to the single pressure design needed for the new burners. Zeeco's Model Simulation Engineering Team spent significant time on-site with the Plant Owner's personnel designing the new system. Zeeco's engineers designed and constructed a scaled Plexiglas model of the entire combustion system and ran the Physical Flow Simulation model in our Connecticut facility. The completed model ensured +/- 3% distribution to each burner with less than 10% peripheral distribution around each burner's air entry.

Figure 6: Free-Jet Physical Flow Simulation

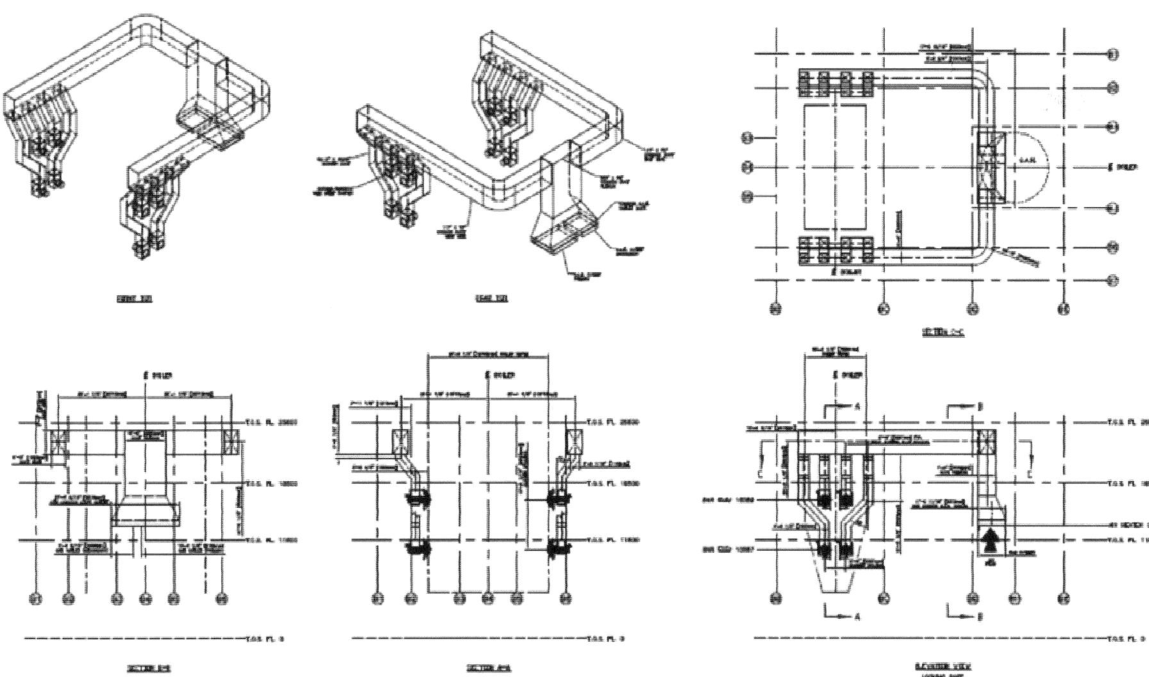

Figure 7: Zeeco's New Design Combustion Air System

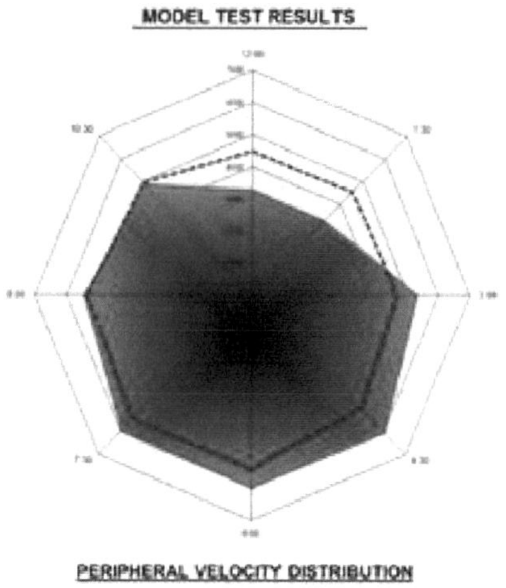

Figure 8: Example of Model Study Peripheral Distribution Chart

Due to the boiler's original controls and Burner Management System's design for main fuel petcoke firing and natural gas start up burners, they were not designed for full load natural gas firing. During the initial design proposal phase, it was agreed that a separate Redundant PLC BMS system would be necessary to integrate into the main DCS system. This would allow the main system to remain "as-is" in the event the unit was to return to petcoke firing in the future. Essentially, it would be a "fuel selector" switch to return the boiler's operation back to petcoke firing. This system was designed in working with the plant owner's original DCS and BMS supplier to ensure a smooth integration into the DCS system.

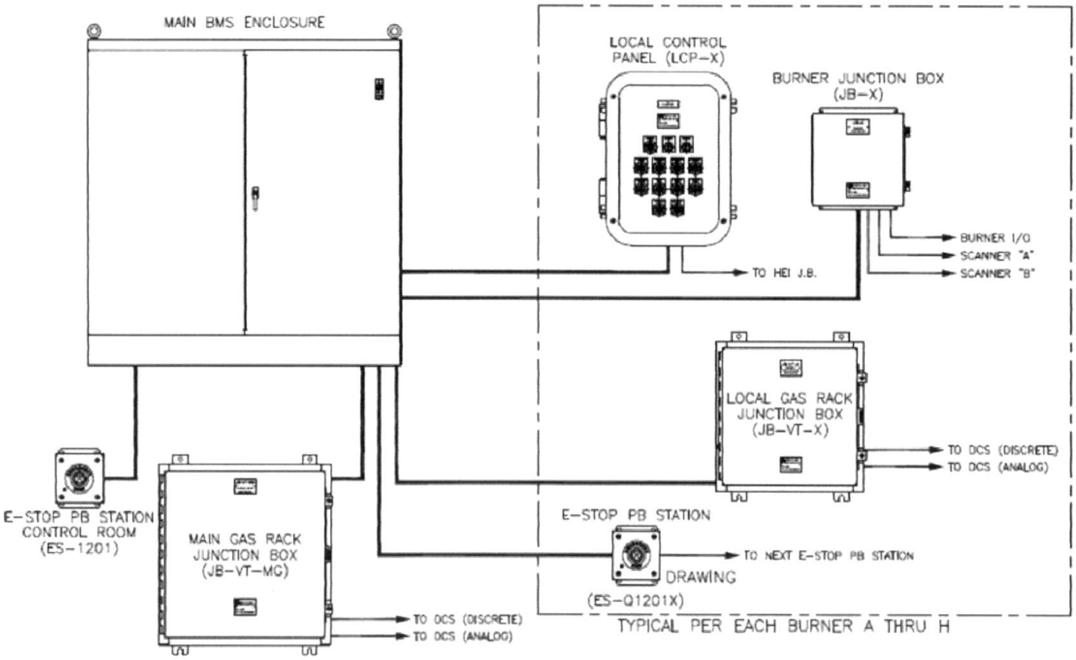

Figure 9: Boiler's Burner Management System

In order for the new burner system to be modified to fire on natural gas, several modifications were required for successful full load natural gas firing, including:

- Removal of furnace's lower slope refractory- During petcoke firing, the turbulent nature of the fluid bed, including the solids return from the cyclone, required the lower slope tubing to be fully protected from potential erosion. When the solid fuel is removed, this lower slope no longer required the refractory coverage to protect the tubing from solids within the cyclone, promoting further heat absorption in the furnace for natural gas firing, and to assist with lower emissions.

- As noted above, the cyclone design in a CFB is to remove solids from the furnace exit flue gas and return them to the fluid bed to promote high efficiency firing and lower carbon loss. With petcoke firing removed, the cyclone was no longer needed for this operation. The inlet to the cyclone plated off, effectively "closing" the inlet so no flue gas would pass through the cyclone. The loop seal exit was "plugged" with refractory to protect it from furnace gasses. This sealing can be deemed 'temporary", as the plate and refractory plug can be removed in the event petcoke firing is deemed cost effective over natural gas firing.

- All back-end clean up equipment are placed out of service due to natural gas firing.

Upon all final design "freeze" and plant owner approvals, fabrication and procurement proceeded and all equipment was delivered to the site by the end of March 2013. Due to the critical schedule, the plant owner was required to take the boiler out of service in late January 2013 to begin the modification process for the natural gas firing. The equipment, upon on-site arrival, was immediately transported to the boiler and the installation process began.

Due to the diligence of the entire engineering team, the integration and installation of the burners, valving and BMS proceeded very smoothly throughout the outage. All the upfront engineering and time ensured the design was cost effective and the outage was completed on time.

Commissioning and Startup

The plant owner also contracted Zeeco's Service Engineers to provide full commissioning and installation assistance, to ensure that all mechanical and electrical equipment was installed per Zeeco's requirements. Zeeco's Service Engineering team was on-site to review and check out the electrical/mechanical/BMS/DCS installation. This proceeded very smoothly during checkout, and all minor issues were addressed in the time allotted for commissioning. Actual equipment start-up proceeded immediately after the final commissioning. Burner performance and full steam boiler operation was achieved in less than two weeks.

Final Conditions are as noted:

- Final emissions performance met the contract and permit requirements of 0.07lbs/MMBtu
- Full load steam boiler requirements were met within the two week timeframe

Conclusion

Boiler steam load and emissions were met in a very timely fashion, and the plant owner was able to generate plant power within less than a month after the initial boiler start-up. The plant owner operated the boiler and turbine/generator near design capacity for over three months after final commissioning. The effort placed on the entire project, pre and post-award, led to the overall success of the boiler's conversion. The integration by the plant owner and the three primary engineering parties led to a fully integrated design that was reviewed by all to ensure the timely success of the project. Based upon the success of the first boiler, Zeeco was awarded the contract for the second CFB boiler conversion, and will begin with installation and commissioning to be in mid-2014.

351797 Vapor Intrusion - Models and Their Value

Abstract:

Vapor Intrusion- Models and Their Value

The problem of vapor intrusion into structures built atop contaminated properties has received increasing attention over the last decade. On the national level, new guidance from the US EPA is imminent, but in the meantime different states have adopted a wide range of policies to address the problem. Many approaches to characterization are empirical, based upon measurements at impacted locations. Since these can involve alarming and intrusive measures involving many residents, some emphasis has been placed on screening, using simple mathematical descriptions of the problem. With the vastly increased availability of powerful fluid dynamics codes, these models can guide more of the site investigation work, as well as evaluate possible possible mitigation strategies. Examples will be given of how such tools can be used to assess both petroleum-based and chlorinated solvent-based situations, based upon recent research carried out in Brown University's Superfund Research Program.

The approach involves solving the equations of vapor transport in the subsurface below the structure of interest, which is typically the soil of the vadose zone. The transport is dominated by diffusion, though entry into the structure may be strongly influenced by local advective transport near the building foundation, as a result of small pressure differences between ambient and the indoors. The indoor environment is typically modeled using a CSTR model. Subtle factors can have a very large influence on predicted indoor air exposures. These include effects related to the resistance of a capillary zone in the case of a chlorinated solvent source in the groundwater, or the form of the assumed biodegradation rate law in the case of petroleum contaminants. The models point to what might represent reliable or problematic field measurement strategies. Also, once a reliable site model is developed, the efficacy of mitigation methods such as subslab depressurization, may be more reliably examined.

351798 Synthesis, Fine Structural Characterization, and CO_2 Adsorption Capacity of Cobalt and Nickel-Based Metal Organic Framework-74

Wednesday, April 2, 2014: 2:30 PM
Newberry (Hilton New Orleans Riverside)

Abhijit Krishna Adhikari, Department of Chemical Engineering & Materials Science/Environmental Technology Research Center, Yuan Ze University, Chung-Li, Taiwan and Kuen-Song Lin, Department of Chemical Engineering & Materials Science/Environmental Technology Research Center, Yuan Ze university, Chung-Li, Taiwan

Abstract: Two metal organic frameworks of MOF-74 group (cobalt and nickel-based) were synthesized, characterized, and evaluated for CO_2 adsorption. The BET specific surface areas of MOF-74(Co) and MOF-74(Ni) MOFs were 1,406 and 1,418 m^2g^{-1}, respectively with the identical pore volume of 0.84 cm^3g^{-1}. CO_2 adsorption capacities of MOF-74(Co) and MOF-74(Ni) were 3.35 and 4.07 $mmol{\cdot}g^{-1}$, respectively measured at 273 K and 1.1 bar. The oxidation state of central atoms in MOF-74(Co) was composed of Co(II) and Co(III) confirmed by XANES spectra while MOF-74(Ni) was Ni(II) central atoms. The bond distances of Co„ŸO and Ni„ŸO were 1.95 and 1.97 Å, respectively.

1. Introduction
Global population and the industrialization are increasing and hence, the energy consumption is also increasing swiftly. Till now, the main source of energy is the burning of fossil fuels. However, burning of fossil fuels releases a vast amount of CO_2 greenhouse gas into atmosphere which causes the raise of average surface temperature of the earth. As the present developments in CCS technology are not sufficiently effective by means of cost and energy consumption, it has prompted to improve the existing technologies and find some new alternatives [1]. Metal organic frameworks (MOFs) are a new class of materials that have attracted substantial attention in past

few years [2]. MOFs are synthesized using organic linkers and metal joints that self-assemble to form different dimensional networks. It can generate unusually large diameter channels and cavities which is particularly attractive for CO_2 gas storage application. Here we report the CO_2 gas adsorption capacity on same type of two MOFs; such as MOF-74(Co) and MOF-74(Ni).

2. Experimental

In a typical experiment, 0.75 g of 2,5-dihydroxyterephthalic acid (DHTA) and 3.0 g of $Co(NO_3)_2{\cdot}6H_2O$ or $Ni(NO_3)_2{\cdot}6H_2O$ were dissolved in 150 mL of DMF with sonication in a 250 mL Duran bottle. The bottle was capped tightly and placed in a 110°C oven for 20 h. After cooling to room temperature, the mother liquor was decanted and the crystals were washed with DMF. The product was immerged into methanol for six days. The products were then evacuated and heated under vacuum to150°C over a period of 12 h.

3. Results and discussion

The FE-SEM images of MOF-74(Co) and MOF-74(Ni) samples are shown in Fig.1a and 1b, respectively. It shows that MOF-74(Co) has cauliflower type structure with the particle size of about 4-6 mm and MOF-74(Ni) has uniformly distributed spherical shape with the average diameter of 2-4 mm, although some particles are aggregated. The XRD patterns (Fig.1c) shows that both samples have the unique crystalline phase indexed on the basis of a face-centered hexagonal unit cell of MOF-74. Both samples indicate almost similar diffraction patterns at identical 2*fÆ* values ascribed the formation of similar crystallographic phases.

Figure 1. -SEM images of (a) MOF-74(Co), (b) MOF-74(Ni), and (c) XRD patterns of synthesized MOF-74 samples.

MOF-74(Co) was thermally stable up to 258°C while in case of MOF-74(Ni) the value was 344°C confirmed from the TGA analysis shown in Fig. 2. The thermogravimetric curves of both samples are consistent with prominent weight-loss steps. It can be seen that the thermal stability is not similar for these two samples although they follow the same trend of loosing weight. In case of MOF-74(Co), a fully hydrated molecular sieve contains about 27 wt.% of water and other guest molecules. and in case of MOF-74(Ni) this value is almost similar, but there is one additional step of weight loss of about 13 wt%.

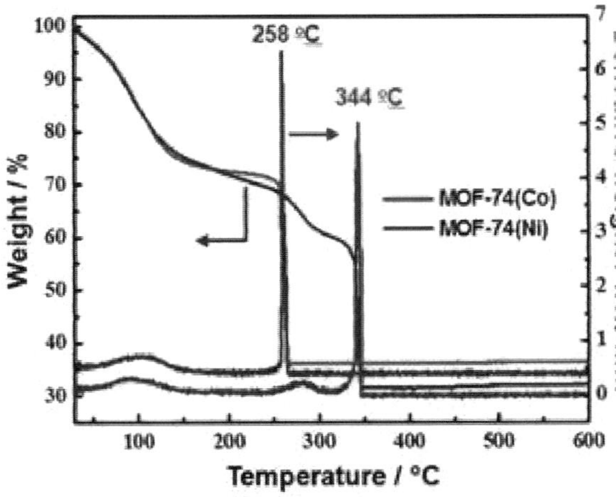

Figure 2. - TG/DTA curves of as-synthesized MOF-74(Co) and MOF-74(Ni)

From the N_2 adsorption isotherms it has been evaluated that both sample follow type-I isotherms with almost no hysteresis loop which corresponding to microporous material. The BET specific surface areas of MOF-74(Co) and MOF-74(Ni) MOFs were 1,406 and 1,418 m^2g^{-1}, respectively with the identical pore volume of 0.84 cm^3g^{-1}. DFT pore size distribution of MOF-74(Zn) and MOF-74(Cu) are shown in Fig. 4a and Fig. 4b, respectively while the values of average pore size and volume are presented in Table 1. These results suggest that these samples are microporous materials.

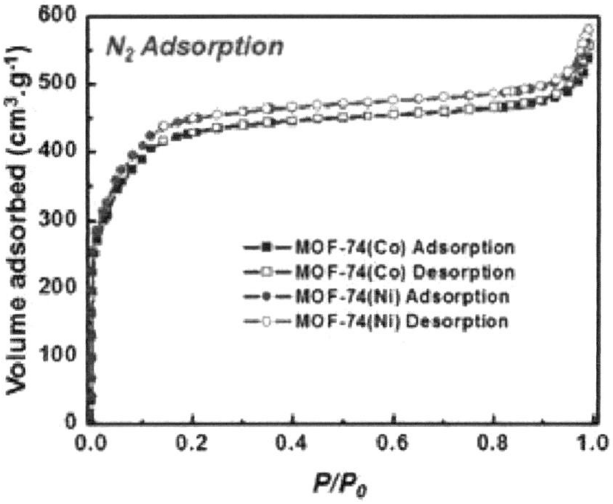

Figure 3. - N_2 adsorption/desorption isotherms of as-synthesized MOF-74(Co) and MOF-74(Ni). Filled and open symbols represent the adsorption and desorption of nitrogen gases, respectively.

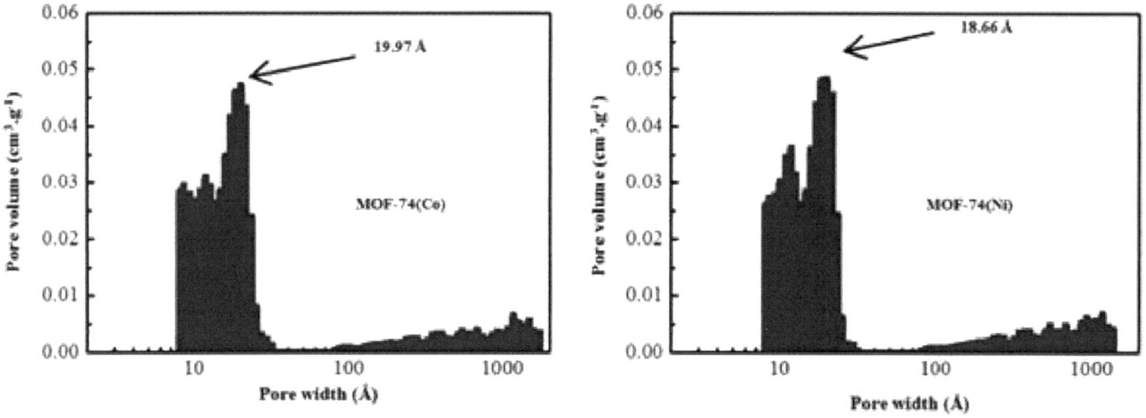

Figure 4. DFT pore size distribution of (a) MOF-74(Co) and (b) MOF-74(Ni)

Table 1. Fine structural parameters of MOF-74 samples analyzed by Co and Ni K-edge EXAFS.

Sample	BET SSA (m^2g^{-1})[a]	Langmuir SSA (m^2 g^{-1})	Pore volume (cm^3g^{-1})[b]
MOF-74(Co)	1,406	2,120	0.87
MOF-74(Ni)	1,418	2,123	0.84

[a] *The specific surface area was measured by BET method; and*
[b] *Pore volume was measured by density functional theory (DFT) method.*

Fine structure and oxidation state of Zn and Cu species in MOF-74 were further investigated through X-ray absorption spectroscopy. The data were collected several times and standard deviation was also calculated. The XANES region of the X-ray absorption spectrum is very informative on both the oxidation and coordination states of metal ions. XANES spectra of MOF-74(Co) and MOF-74(Ni) samples are shown in Fig. 5a and 5b, respectively along with the standard materials. The comparison is designed to determine the oxidation states of center atoms in MOFs and understanding the effects of DHTA linkers and guest molecules on the crystal structures. It was found that the edge position (7709.46 eV) of as-synthesized MOF-74(Co) sample is very close to the Co(II). However, compared with the standard Co(II) the offset was +0.46 eV, revealed that Co ions were not entirely surrounded by oxygen atoms and hence formed hydroxides (OH^-) through covalent bond. This result reveals that the as-synthesized sample was not totally moisture or guest molecule free. The source of the OH^- ions is either adsorbed moisture or the organic linkers that cause the little distortion in the crystal structure. In case of MOF-74(Ni) the offset was +0.57 eV. Fine structural parameters for both samples are shown in Table 2 where it can be seen that the bond distances of Co„ŸO and Ni„ŸO were 1.95 and 1.97Å, with the coordination number of 4.27 and 4.23, respectively.

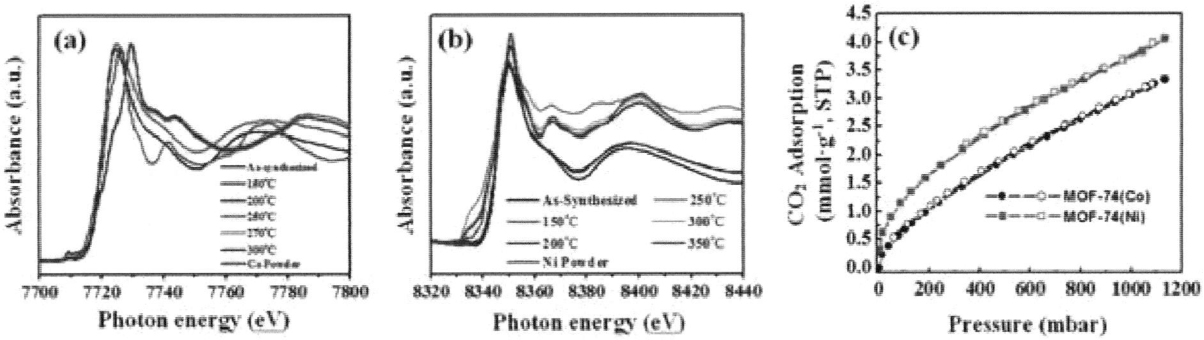

Figure 5. (a) Co K-edge XANES spectra of MOF-74(Co) samples with Co(0) standard, (b) Ni K-edge XANES spectra of MOF-74(Ni) with Ni(0) standard, and (c) CO_2 adsorption isotherms of MOF-74 samples measured at 273 K. Filled and open circles represent the adsorption and desorption, respectively.

Table 2. Fine structural parameters of MOF-74 samples analyzed by Co and Ni K-edge EXAFS.

Sample	Shell	CN[a] ± 0.5	R[b] ± 0.002 (Å)	$\Delta\sigma^{2c}$ (Å^2)
MOF-74(Co)	Co—O	4.27	1.953	0.0037
MOF-74(Ni)	Ni—O	4.23	1.974	0.0038

[a]*Coordination number;* [b]*Bond distance;* [c]*Debye-Waller factor.*

The CO_2 adsorption capacities of the synthesized samples are shown in Fig. 5c. The steady ascending slope of the CO_2 adsorption isotherms represents strong adsorption of CO_2 molecules having quadrupoles onto the open metal sites in MOF-74 samples. The CO_2 adsorption capacity of MOF-74(Ni) measured at 273 K and 1.1 bar was 4.07 mmol·g^{-1}; significantly higher than MOF-74(Co) which was 3.35 mmol·g^{-1}. However, both results are much higher than other prospective adsorbents of CO_2 such as Zeolite 13X [3] and ZIF-96 [4]. The elevation in CO_2 adsorption in MOF-74(Ni) may attribute to the higher surface area than MOF-74(Cu) and more importantly the strong affinity to CO_2 molecules due to different electronic configurations. It can be seen that the adsorption and desorption cycles are almost reversible for both samples. This integrity was also corroborated by the N_2 adsorption. The steady uphill slope of the adsorption curves indicates that higher adsorption capacity may obtain at higher pressure.

4. Conclusions

The MOF-74 samples were capable to store significant amount of CO_2 at relatively lower pressure. The difference in CO_2 adsorbed amount between two samples may attribute to the different affinity of cobalt and nickel ions to CO_2 molecules as well as the specific surface area. The moderate thermal stability confirms that these MOFs may apply to control the CO_2 gas from the emission of refinery or other industries.

References

1. A. Torvanger, M. T. Lund, N. Rive, Mitig. Adapt. Strat. Gl. 18 (2013) 187.
2. T. G. Glover, G. W. Peterson, B. J. Schindler, D. Britt, O. M. Yaghi, Chem. Eng. Sci. 66 (2011) 163.
3. J.-S. Lee, J.-H. Kim, J.-T. Kim, J.-K. Suh, J.-M. Lee, C.-H. Lee, J. Chem. Eng. Data 47 (2002) 1237.
4. W. Morris, B. Leung, H. Furukawa, O. K. Yaghi, N. He, H. Hayashi, Y. Houndonougbo, M. Asta, B. B. Laird, O. M. Yaghi, J. Am. Chem. Soc. 132 (2010) 11006.

Treatment of Flowback Water from Hydraulic Fracturing with Biochar

Maoqi (Mark) Feng

Steven Cooks

Zhigang Feng

April 2, 2014

AIChE 2014 Spring Conference, New Orleans

Chemical Engineering Dept., Div. of Chemistry and Chemical Engineering, Southwest Research Institute, San Antonio, TX 78238

Outline

- Introduction
- Technical Approach
- Experimental Work
- Summary & Conclusions

2

Water Demand in Hydraulic Fracturing

- Water demand is huge: 100 billion gallons of water used for 39,000 wells in 2011
- Fracking accounts for 50% of total water consumption in some Texas areas

3

Water Requirement in Shale Gas Production

- Water requirements for hydraulic fracturing, by shale plays in Texas

Water Requirements (gallons per well)	Low Value	Median Value	High Value
Barnett Shale	<1 MM	2.6 MM	>8 MM
Haynesville and Bossier Shale	< 1 MM	5.5-6 MM	>10 MM
Eagle Ford Shale	1 MM	6-6.5 MM	13 MM
Woodford, Pearsall, and Barnett-PB Shale	<1 MM	0.75-1 MM	<5 MM

4

Flowback Water Recycling

- Half of fracturing wells are in water-stressed areas
- Flowback water reuse is needed

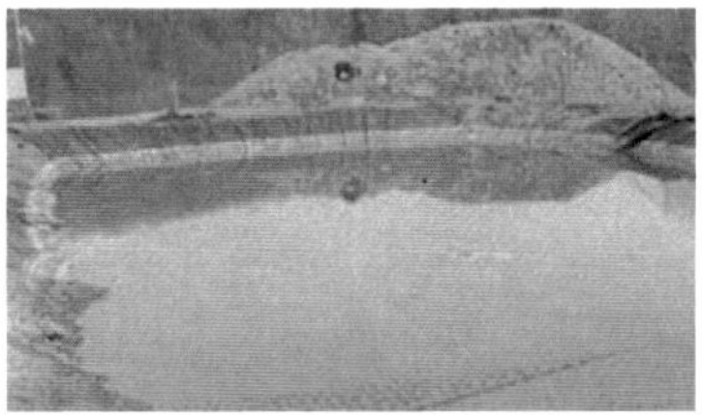

5

Reusable Water Quality

- Parameter Range

TDS at 180°C, mg/l	9,000–16,000
Turbidity, NTU	0-5
pH	6.5-8
Iron, mg/l	1-10
Chloride, mg/l	5,000-10,000
Potassium, mg/l	100-500
Calcium, mg/l	50-250
Magnesium, mg/l	10-100
Sodium, mg/l	2,000-5,000
Boron, mg/l	0-20

6

Flowback Water Treatment

- Membrane method is expensive and cannot handle solids and hydrocarbons
- Zeolites generate more solid waste
- Bioremediation is not efficient
- Nanofiltration is easily plugged
- Activated carbon adsorption is more expensive

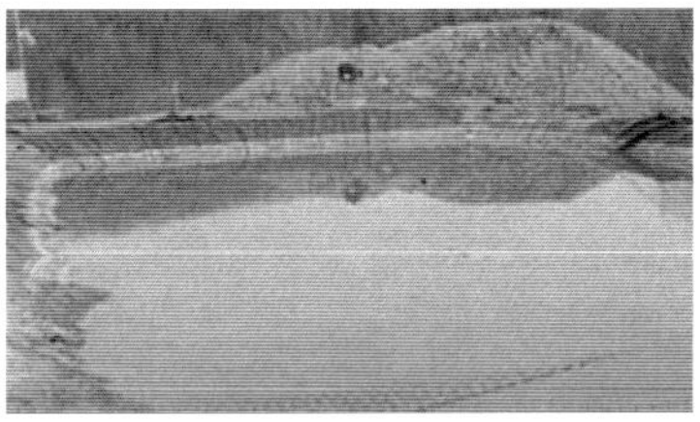

7

Why Biochar

- Biochar is a good sorbent
- Green
- Biodegradable
- Soil compatible and amendment
- Easy to prodcue
- cost effective
- Mass production available

8

Biochar vs. Activated Carbon

- Biochar is produced at lower temperature
- Biochar is a precursor for activated carbon
- Activated carbon needs an activation step
- Biochar is more feedstock flexible

9

Biochar Production

- Biochar production from wood chips by pyrolysis
- Temperature 350°C – 550°C

10

Biochar Property

- Pine wood biochar carbon content vs. pyrolysis temperature

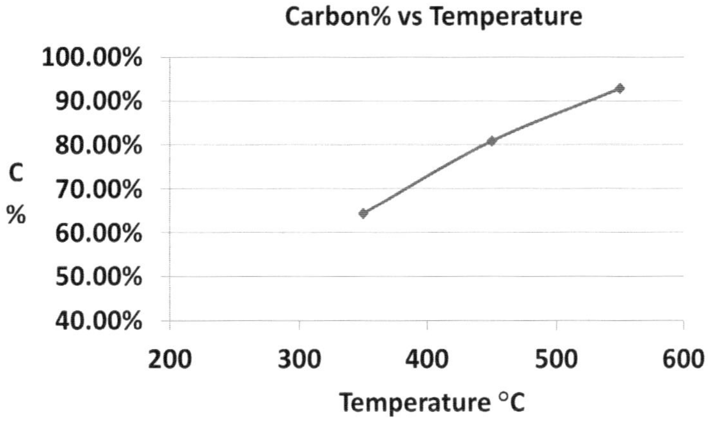

11

Biochar Property

- Oak wood biochar carbon content vs. pyrolysis temperature

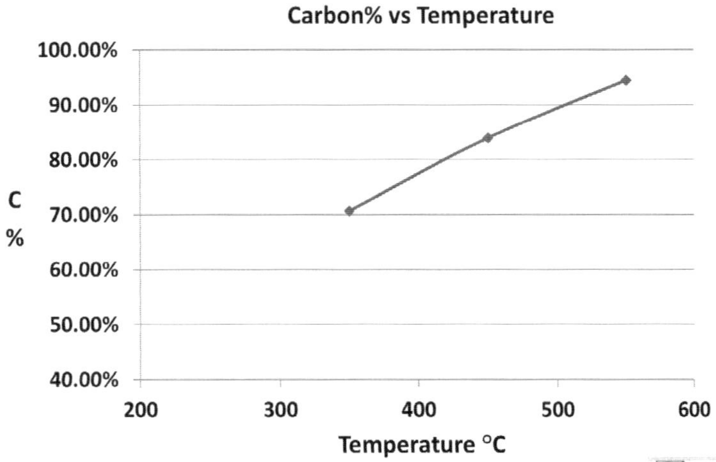

12

Biochar Preparation

- Biochar prepared from different sources at 350°C

Feed	Carbon%
Cellulose	90.09%
Pine Wood	66.43%
Oat Hulls	58.59%

13

Batch Adsorption

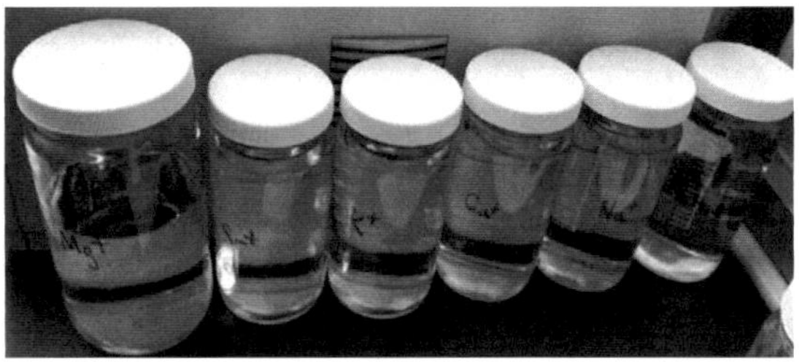

14

Batch Adsorption

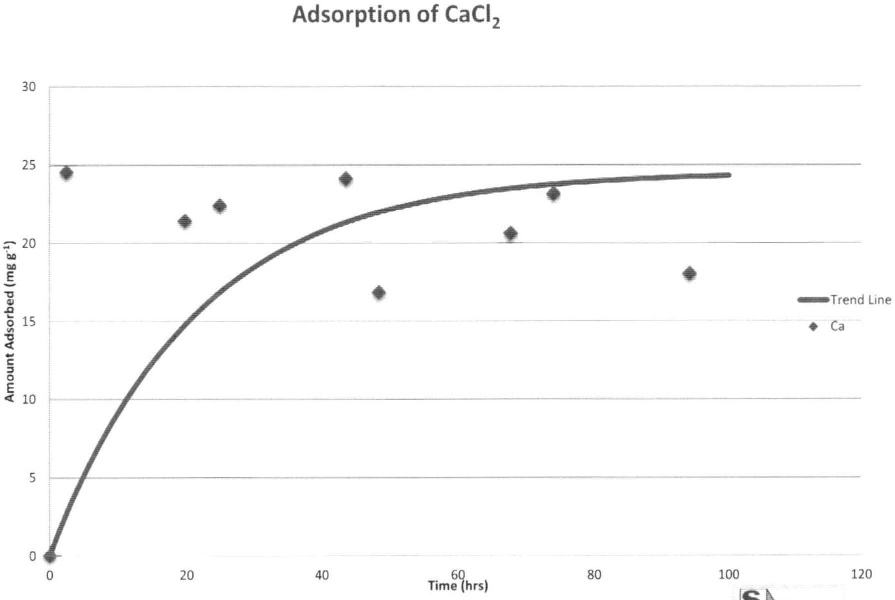

15

Sorbent Bed Evaluation in Continuous Flow Mode

- Quick evaluation of sorbent capacity in continuous flow mode – more efficient than batch mode

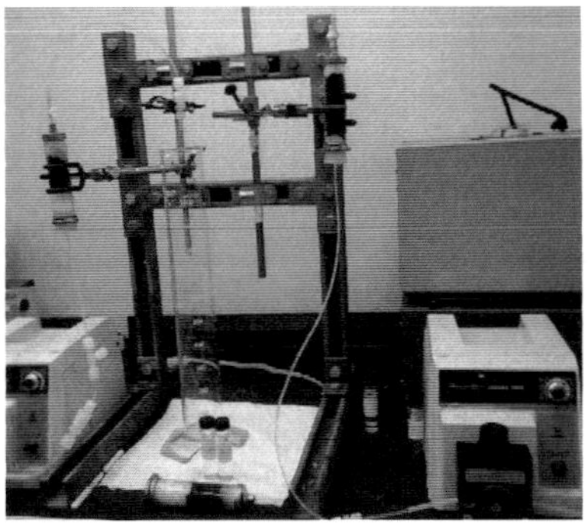

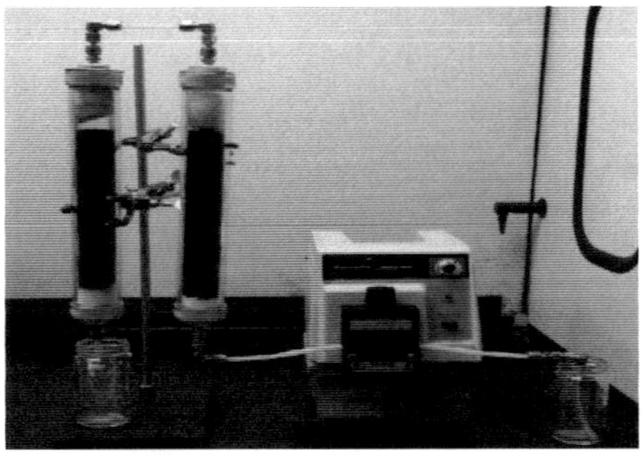

16

NaCl Removal

- NaCl removal

Time (hrs)	NaCl Mass (g)	Water Mass (g)	NaCl Concentration, wt%	NaCl wt% Adsorbed
0.00	12.75	247.5	4.90%	0.00%
2.62	0.8921	18.5788	4.58%	0.32%
19.53	0.9598	19.3488	4.73%	0.17%
24.57	0.9886	19.7544	4.77%	0.13%
43.37	0.9908	20.1408	4.69%	0.21%
48.32	0.9065	18.6041	4.65%	0.25%

17

Mg Removal

- $MgCl_2$ removal

Time (hrs)	$MgCl_2$ Mass (g)	Water Mass (g)	$MgCl_2$ Concentration, wt%	$MgCl_2$ wt% Adsorbed
0.00	4.8754	38.5514	11.23%	0.00%
0.5	4.1715	38.4528	9.79%	1.44%
1.0	4.3688	39.0205	10.07%	1.16%
1.5	4.6866	39.086	10.71%	0.52%
2.0	4.3229	38.8608	10.01%	1.22%

18

Ca Removal

- $CaCl_2$ removal

Time (hrs)	$CaCl_2$ Mass (g)	Water Mass (g)	$CaCl_2$ Concentration, wt%	$CaCl_2$ wt% Adsorbed
0.00	3.975	38.7583	9.30%	0.00%
0.5	3.9614	39.2739	9.16%	0.14%
1.0	3.936	38.9826	9.17%	0.13%
1.5	3.9684	39.5214	9.12%	0.18%
2.0	3.9067	39.0601	9.09%	0.21%
2.5	3.9021	39.369	9.02%	0.28%

19

Sr Removal

- $Sr(NO_3)_2$ removal

Time (hrs)	$Sr(NO_3)_2$ Mass (g)	Water Mass (g)	$Sr(NO_3)_2$ Concentration, wt%	$Sr(NO_3)_2$ wt% Adsorbed
0.0	0.4834	20.7043	2.28%	0.00%
0.5	0.4557	20.665	2.16%	0.12%
1.0	0.4491	21.1634	2.08%	0.20%
1.5	0.4191	20.3263	2.02%	0.26%
2.0	0.4114	20.5186	1.97%	0.32%
2.5	0.4031	20.1342	1.96%	0.32%

20

Inorganics Removal

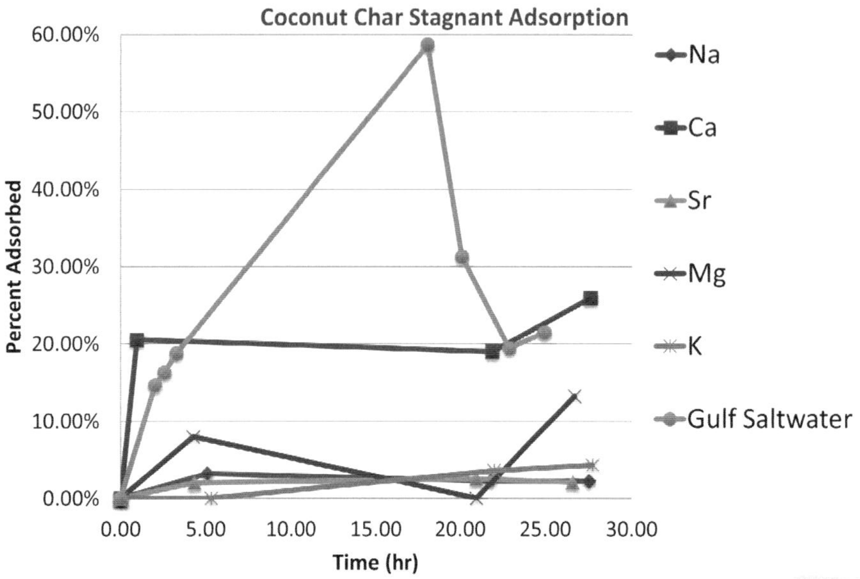

21

Hydrocarbons Removal

- Removal of diesel hydrocarbons in water by biochar adsorption
- Per pass removal 87%
- Higher adsorption capacity for biochar produced at higher temperature

22

Solid Particulates Removal

- Removal of solid particulates in flowback is necessary
- Pump damage
- Long settling time
- Fine solids removal is still in progress

23

Summary

- Biochar production from wood chips
- Quick evaluation of sorbent capacity in continuous flow mode
- Successful removal of inorganic salts: NaCl, K^+, $MgCl_2$, $CaCl_2$, Sr^{2+}
- Successful inorganics removal from sea water
- Successful removal of hydrocarbons from simulated flowback water

24

Some Thoughts

- Biochar adsorption bed can be used with other methods, such as ultrafiltration, reverse osmosis, and ion exchange units
- Biochar sorbent bed can be used as a pretreatment or post-polishing step for other methods

25

Acknowledgement

Thank you!

Questions?

26

352927 Hazards Management for Building Locations through Air Dispersion Modeling

Abstract:

Oil, gas, and petrochemical industries have reached a critical era involving its timeworn infrastructure. Previously, facilities degraded with minimal effort allocated for updates that ensure personnel safety (i.e., process inventory isolation, automative control systems, etc.). More recently, industry as a whole is progressively adopting a more safety conscientious attitude about personnel and the environment. These personnel and environmental concerns are driving new codes, standards, recommended practices, and regulations to be developed and implemented. Safety studies are being performed which assess facilities to determine if these new regulations are met, while providing key insight to eliminate any deficiencies that are discovered. Available advanced air dispersion models estimate or predict the concentration of air pollutants or toxicants emitted from various sources such as industrial plants, road traffic vehicles or accidental chemical releases. Facility siting assessments are amongst the required studies used for managing the risk to on-site personnel located in new and existing buildings intended for occupancy, where air dispersion models can be used to determine the consequences of accidental releases of hazardous or toxic materials. This paper discusses case studies carried out for hazards management for building locations with different air dispersion modeling software tools including: SCREEN3, AERMOD, ALOHA, and PHAST.

353139 H2S Removal from Biogas Using an Ionic Liquid as Physical Absorbent

Abstract:

Biogas is nowadays a valuable renewable energy carrier which can be used directly as a fuel or as a raw material for the production of synthesis gas and/or hydrogen. Methane (CH_4) and carbon dioxide (CO_2) are the main constituents, but biogases also contain significant quantities of undesirable compounds (contaminants), such as hydrogen sulfide (H_2S) and ammonia (NH_3). The existence and quantities of these contaminants depend on the biogas source (i.e., landfills, anaerobic fermentation of manure). Their presence constitutes a major problem because (1) they can adversaly affect any biogas thermal or thermocatalytic conversion device (e.g., corrosion, erosion, fouling); and (2) they produce harmful environmental emissions. It is therefore important to develop a reliable and efficient biogas purification process. This work is aimed at presenting an alternative removal approach for H_2S from biogas based on physical absorption using an ionic liquid as solvent.

The Growth of Idea Ownership: Patent Infringement and Risk Management within the Chemical Process Industries

Paul LaVanway, Esq.
Fredrikson & Byron, P.A.

14 AIChE
Spring Meeting, New Orleans, LA

Fredrikson & BYRON, P.A.

Overview of Presentation

I. Patent Activity Update

II. Practical Risk Management Techniques

III. New Tools

14 AIChE
Spring Meeting, New Orleans, LA

Fredrikson & BYRON, P.A.

PATENT ACTIVITY UPDATE

14AIChE®
Spring Meeting, New Orleans, LA

Fredrikson & BYRON, P.A.

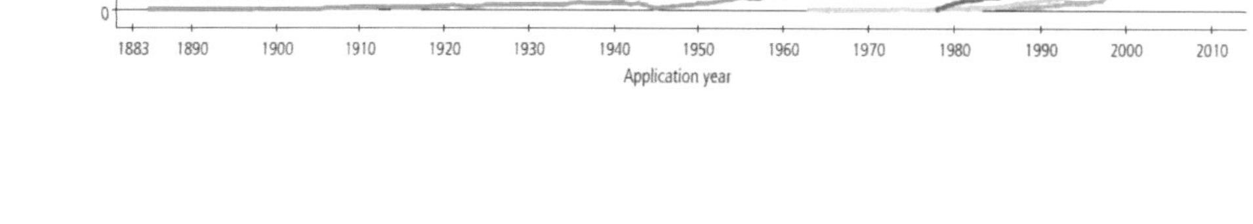

14AIChE®
Spring Meeting, New Orleans, LA

Fredrikson & BYRON, P.A.

Worldwide Filing Trends

Origin	Australia	Canada	China	Finland	France	Germany	Italy	Japan	Netherlands	Republic of Korea	Russian Federation	Sweden	Switzerland	United Kingdom	United States of America	Others
	Field of technology															
Chemistry																
Organic fine chemistry	1,164	2,798	18,900	539	20,317	38,897	4,436	38,136	6,148	10,424	2,185	5,375	17,365	12,314	75,204	30,237
Biotechnology	2,592	3,432	17,402	804	8,107	16,915	2,306	20,814	5,500	9,165	1,822	1,765	7,316	7,058	67,521	21,840
Pharmaceuticals	3,647	6,297	38,194	958	18,328	31,204	7,370	28,798	6,362	10,514	5,692	7,748	23,615	15,563	123,173	50,208
Macromolecular chemistry, polymers	382	816	11,135	2,173	4,439	20,009	2,446	45,323	4,576	7,906	935	325	3,439	1,714	31,810	7,538
Food chemistry	875	1,122	19,191	428	2,595	5,330	1,070	13,272	6,409	12,111	17,015	337	4,744	2,464	19,963	11,149
Basic materials chemistry	1,002	1,992	24,266	901	6,012	35,527	1,536	42,724	7,038	11,113	3,343	690	6,970	6,340	51,100	13,535
Materials, metallurgy	1,697	1,569	30,285	1,729	6,697	17,150	1,480	43,685	1,846	12,235	8,359	1,514	2,186	2,422	20,753	12,182
Surface technology, coating	670	1,386	11,959	1,185	4,907	15,954	1,642	52,591	1,933	10,246	1,779	1,200	2,604	2,498	36,858	7,336
Micro-structural and nano-technology	116	96	1,710	124	606	1,371	94	2,405	254	2,579	587	134	153	158	2,486	683
Chemical engineering	1,479	2,235	16,952	1,840	7,352	25,809	2,846	32,504	4,788	13,354	4,426	2,422	4,287	5,490	40,121	13,512
Environmental technology	762	1,514	13,620	855	5,042	13,594	1,481	27,259	2,533	12,888	2,303	1,116	1,765	2,719	20,263	8,002

14 AIChE® Spring Meeting, New Orleans, LA

Fredrikson & BYRON, P.A.

Worldwide Filing Trends

Field of technology	Publication year					Growth rate
	2007	2008	2009	2010	2011	2007-11 (%)
Chemistry						
Organic fine chemistry	52,609	53,873	53,009	52,559	51,271	-0.6
Biotechnology	33,929	35,654	37,652	38,423	40,849	4.7
Pharmaceuticals	73,345	73,769	72,126	69,258	69,311	-1.4
Macromolecular chemistry, polymers	27,627	28,345	28,800	28,683	28,748	1.0
Food chemistry	21,690	23,582	27,245	27,889	30,858	9.2
Basic materials chemistry	38,920	41,085	42,322	43,795	45,115	3.8
Materials, metallurgy	30,200	33,775	34,968	36,988	38,542	6.3
Surface technology, coating	29,348	30,732	32,792	33,215	33,954	3.7
Micro-structural and nano-technology	2,302	2,478	2,829	3,094	3,203	8.6
Chemical engineering	33,263	35,087	35,766	36,618	37,991	3.4
Environmental technology	21,688	22,605	24,333	25,572	26,324	5.0

14 AIChE® Spring Meeting, New Orleans, LA

Fredrikson & BYRON, P.A.

Comparative Trends

Field of technology	Publication year					Growth rate
	2007	2008	2009	2010	2011	2007-11 (%)
Electrical engineering						
Electrical machinery, apparatus, energy	101,013	105,591	111,876	116,009	122,697	5.0
Audio-visual technology	92,978	91,256	85,411	80,607	75,881	-5.0
Telecommunications	68,915	69,804	60,808	56,417	49,533	-7.9
Digital communication	57,696	64,237	69,191	74,389	79,726	8.4
Basic communication processes	17,208	17,734	17,164	16,515	15,554	-2.5
Computer technology	123,504	134,493	133,072	130,052	134,272	2.1
IT methods for management	19,370	21,684	25,022	23,261	23,532	5.0
Semiconductors	76,617	81,238	78,685	77,547	80,049	1.1
Mechanical engineering						
Handling	42,764	42,783	42,764	42,414	44,344	0.9
Machine tools	36,607	38,315	40,585	43,242	46,363	6.1
Engines, pumps, turbines	41,891	43,573	48,156	48,372	48,423	3.7
Textile and paper machines	35,815	33,681	32,251	30,633	30,364	-4.0
Other special machines	44,008	45,912	47,541	49,011	50,981	3.7
Thermal processes and apparatus	25,280	25,699	27,259	29,382	29,855	4.2
Mechanical elements	43,957	47,513	47,286	46,405	46,913	1.6
Transport	65,237	67,717	70,542	67,015	65,439	0.1

14 AIChE Spring Meeting, New Orleans, LA

Fredrikson & BYRON, P.A.

Change in 5 year average over last 20 Years

Fastest Growing

Category	Growth Rate
Computer Software	+11%
Data Processing	+11%
Semiconductor Devices	+10%
Video Distribution	+10%
Computer Hardware	+8%
Chemical – Crystals	+8%
Nanotechnology	+8%
Information Storage	+6%
Communications	+6%
Design	+5%

Slowest Growing

Category	Growth Rate
Chemical – Purification/Evaporation /Distillation	-2%
Chemical – General Compounds	-3%
Time Measurement	-3%
Machine Mechanism	-3%
Chemical Manufacture Specific	-3%
Organic Compounds	-3%
Pipes & Joints	-4%
Education	-4%
Hazardous Waste	-4%
HVAC	-4%

14 AIChE Spring Meeting, New Orleans, LA

Fredrikson & BYRON, P.A.

US Patent Filing Trends

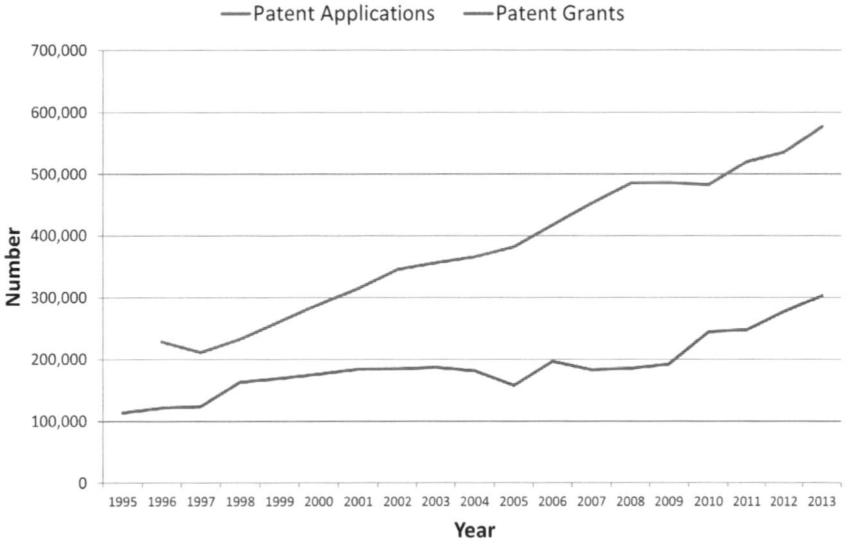

14 AIChE Spring Meeting, New Orleans, LA

Fredrikson & BYRON, P.A.

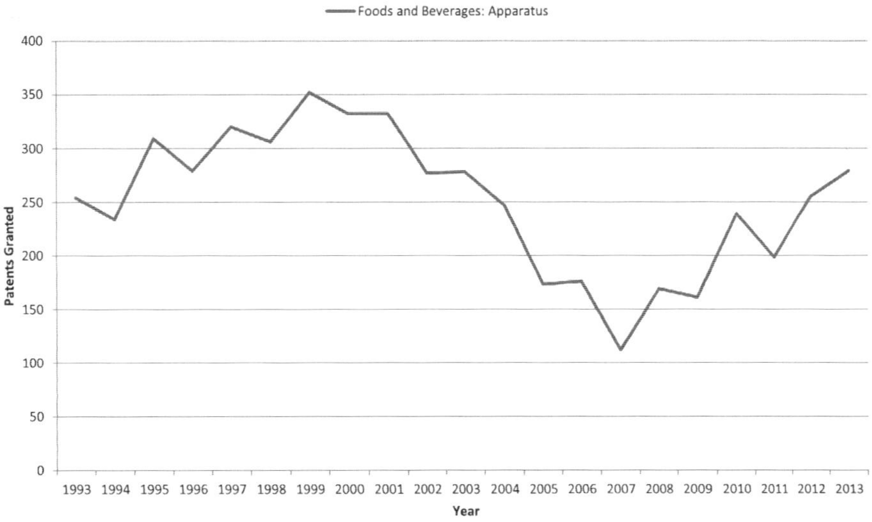

14 AIChE Spring Meeting, New Orleans, LA

Fredrikson & BYRON, P.A.

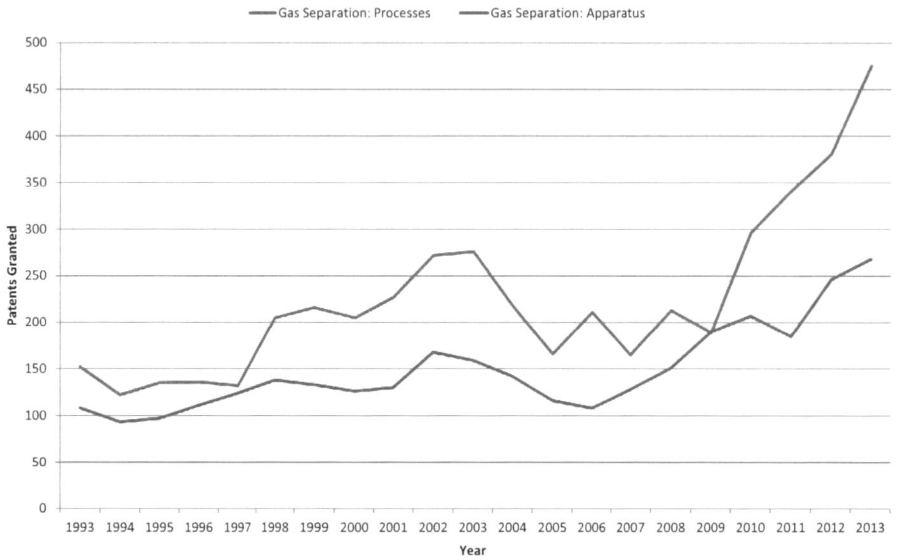

14 AIChE® Spring Meeting, New Orleans, LA

Fredrikson & BYRON, P.A.

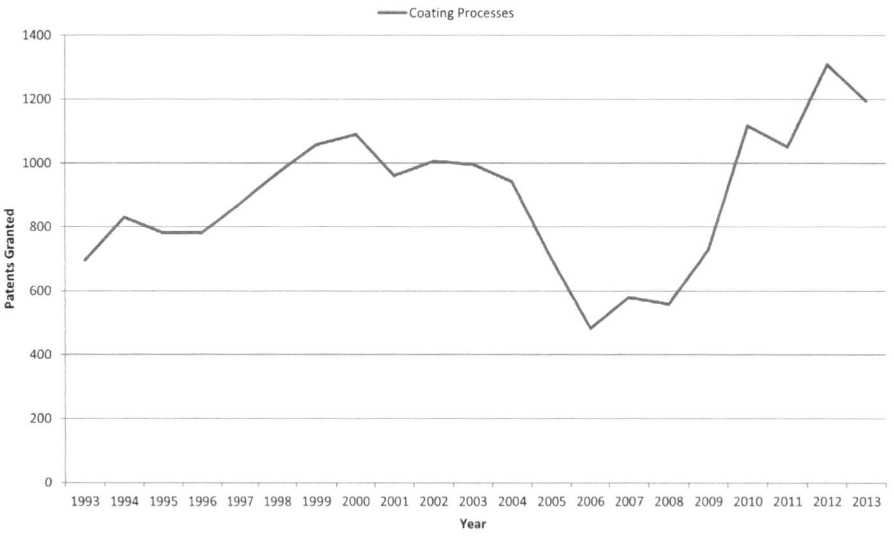

14 AIChE® Spring Meeting, New Orleans, LA

Fredrikson & BYRON, P.A.

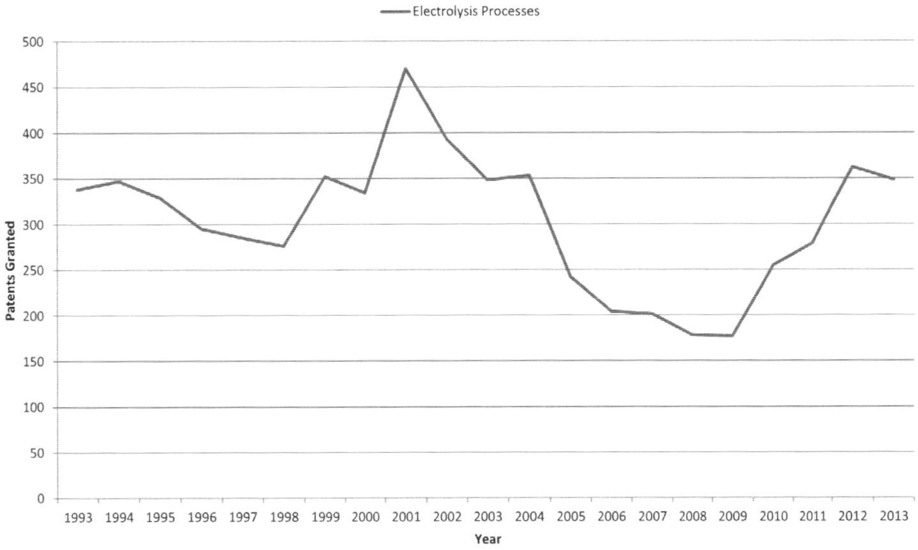

14 AIChE®
Spring Meeting, New Orleans, LA

Fredrikson & BYRON, P.A.

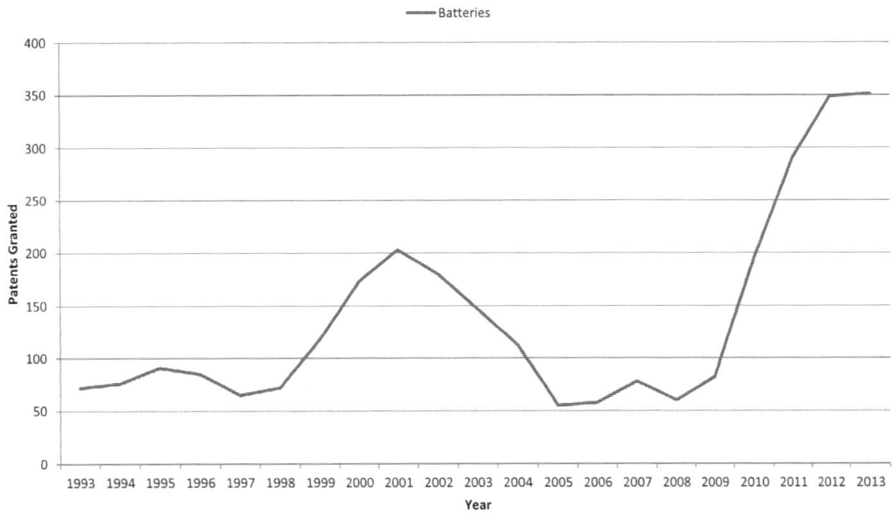

14 AIChE®
Spring Meeting, New Orleans, LA

Fredrikson & BYRON, P.A.

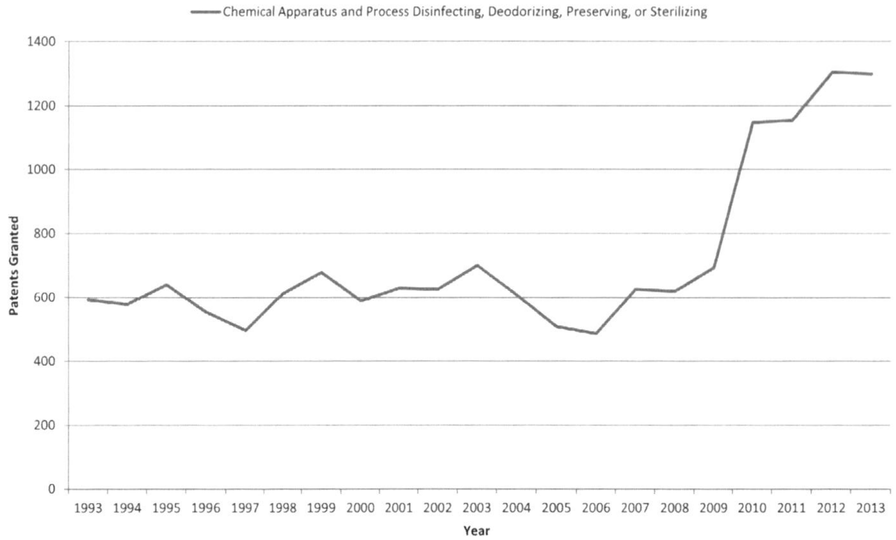

14 AIChE® Spring Meeting, New Orleans, LA

Fredrikson & Byron, P.A.

Patent Litigation Statistics

- Median damage award last 15 years all cases
  - $1.9 M → 16.5 M
- Median over last 5 years
  - $4.9M
- Chemical Industry (excluding bio/pharma)
  - Generally smaller median damages (less than $3M) but characterized by outliers
- Disruptive
  - Median time to trial 2.3 years

14 AIChE® Spring Meeting, New Orleans, LA

Fredrikson & Byron, P.A.

Cost of Patent Infringement Litigation

Amount at Risk	Median Cost
Less than $1 Million	$700,000
$1-$10 Million	$2,000,000
$10-$25 Million	$3,325,000
$25 Million+	$5,500,000

14 AIChE
Spring Meeting, New Orleans, LA

Fredrikson & BYRON, P.A.

Distribution of Patent Cases

Top Ten Industry	Percent of Cases
Consumer Products	18%
Biotechnology / Pharma	13%
Industrial / Construction	11%
Computer Hardware / Electronics	9%
Medical Devices	9%
Business / Consumer Services	6%
Software	6%
Automotive / Transportation	5%
Telecommunications	5%
Chemical / Synthetic Materials	5%

14 AIChE
Spring Meeting, New Orleans, LA

Fredrikson & BYRON, P.A.

But . . . Looking at 11-15

Rank	Industry
11	Food/Beverages/Tobacco
12	Metals/Mining
13	Clothing/Textiles
14	Energy
15	Agriculture

14 AIChE® Spring Meeting, New Orleans, LA

Fredrikson & BYRON, P.A.

RISK MANAGEMENT STRATEGIES

14 AIChE® Spring Meeting, New Orleans, LA

Fredrikson & BYRON, P.A.

Understanding the Nature of the Risk

- "Fuzzy" boundaries Problems

<u>Features</u>

- ~~Back~~
- Seat
- ~~Four~~ Legs resting on the ground

<u>Revised</u>

- Seat
- At least one base contacting the ground

14AIChE Spring Meeting, New Orleans, LA

Fredrikson & BYRON, P.A.

Understanding the Nature of the Risk

<u>Revised</u>

- Seat
- At least one base contacting the ground

Legal Scope Not Necessarily Co-extensive with Thrust of Description

14AIChE Spring Meeting, New Orleans, LA

Fredrikson & BYRON, P.A.

Managing New Product Risk

- Correlate Effort to Risk
  - Repeat Use of Old Technology in Existing Field
  - Old Technology in Different Field
  - New Technology in Existing Field
  - New Technology in New Field

Increasing Risk

14 AIChE Spring Meeting, New Orleans, LA

Fredrikson & BYRON, P.A.

Other Factors to Consider

- Intensity of Competition within Field
- Past History Between Competitors
- Value of Marketplace and Opportunity
- Relative Size of Parties
- Monetization Proposition

14 AIChE Spring Meeting, New Orleans, LA

Fredrikson & BYRON, P.A.

Managing New Product Risk

- Use a Phased Approach
  - Phase I
    - Basic Understanding of Landscape and Any "Show Stopping" Patents
    - Focus on Likely Competitors – Review Patents, Literature, Products
  - Phase II
    - Revisit Base Design and Evaluate Ancillary Features
  - Phase III
    - Check Any Design Modifications or Newly Issuing Patents

14 AIChE®
Spring Meeting, New Orleans, LA

Fredrikson & BYRON, P.A.

Managing Ongoing Risk

- Monitor Key Competitors / Inventors
- Monitor Technology Space
- Work with Sales and Marketing Personnel to Track New Product Introduction
- Watch Partners and Consultants

14 AIChE®
Spring Meeting, New Orleans, LA

Fredrikson & BYRON, P.A.

NEW TECHNIQUES FOR ACTION

14 AIChE
Spring Meeting, New Orleans, LA

Fredrikson & BYRON, P.A.

Knowing What Tools Are Available

- Pre-Issuance Submission
  - Attack the Patent Application Before it Becomes a Problem
- Simple But Limited Patent Office Attack
  - Ex Parte Reexamination
- More Involved Patent Office Attacks
  - Inter Partes Review
  - Post Grant Review

14 AIChE
Spring Meeting, New Orleans, LA

Fredrikson & BYRON, P.A.

CONCLUSIONS

- CPI Patents Generally Trending Upward
- Great Sector Specific Variability
- Effective Risk Management Requires Being Proactive and Deploying Resources Efficiency
- New Tools to Attack Problem Patents

14 AIChE® Spring Meeting, New Orleans, LA

Fredrikson & BYRON, P.A.

QUESTIONS?

14 AIChE® Spring Meeting, New Orleans, LA

Fredrikson & BYRON, P.A.

Chemical Engineering & the Law --Case Examples in Industrial Litigation

Paper# 352190

Satish Almaula
Senior Principal Consultant
Baker Engineering and Risk Consultants, Inc.
salmaula@bakerrisk.com

Presented at American Institute of Chemical Engineers Spring Meeting, New Orleans LA April 2014

Abstract

Chemical, petroleum, and allied manufacturing sectors involved in research, technologies, engineering and project management at times encounter issues and problems in successfully meeting a project's intended financial, technical, and operational goals and expectations. Significant deviations from goals and expectations can result in lawsuits between involved parties. Alternatively, parties file insurance claims, which can also result in lawsuits.

Research and technology issues can result in patent and intellectual property legal cases. In addition, issues can arise from engineering design, licensing, product liability, safety, regulatory compliance, construction and operational deficiencies, or an inability to meet contractual and performance commitments. Accidents and incidents at manufacturing or other facilities can result in damage to property and the environment, injuries and death as well as interruption to normal operation causing business loss. Such damages and losses also can result in lawsuits and insurance claims.

The need for collaboration between lawyers, scientists, engineers and other technical specialists is obvious when conducting legal, technical and business analyses to get to the root cause of the matter to support or defend a legal complaint or an insurance claim.

This paper presents a brief introduction to lawsuits and how legal and technical teams can collaborate with their respective experience and expertise in handling a lawsuit either as a plaintiff's or a defendant's team.

The main focus of the presentation will be case examples that demonstrate how principles familiar to engineers and other technical specialists play a part in litigation involving diverse issues in the chemical, refining and allied business sectors.

1 Introduction

Technical professionals, whether they are engineers, scientists or technical specialists often get drawn into legal disputes. Modern day media reports of claims and settlements tend to focus heavily on events in the electronic, information technology, pharmaceutical and drug industries, especially when the issues have a direct effect on consumers and involve huge sums of money.

Petroleum, chemical processing, and allied manufacturing sectors have their share of claims and disputes involving large amounts of money but do not always get the high visibility as the above-mentioned industries. Catastrophic incidents such as fires, explosions, release of hazardous materials leading to deaths, injuries and property damage do get expanded coverage and high visibility.

It is not surprising, therefore that chemical engineers, scientists and technology specialists are often drawn into insurance claims and legal disputes. Lawyers, insurers, scientists, engineers, and technical specialists often collaborate in such matters.

This presentation provides a brief background on typical issues leading to legal disputes followed by selected case examples. Names of organizations, involved parties, and individuals are not disclosed or discussed in this presentation. The selected examples are from the author's experience in lawsuits related to chemical processing and allied industries.

2 Insurance Claims and Lawsuits

Insurance claims and lawsuits can arise from diverse losses and disputes such as property damage, fatalities and injuries, negligent performance or, in extreme but uncommon instances, they can arise from intentional or criminal acts.

Losses alleged in processing industry claims and disputes tend to be related to: patent and intellectual property; engineering design; licensing; product liability; safety; regulatory compliance; construction and operational deficiency; and the inability to meet contractual and performance commitments.

Technical professionals including those in chemical engineering and similar fields can be drawn into insurance claims and lawsuits as fact or expert witnesses.

Fact witnesses are generally limited to testifying about facts that they know firsthand about an issue without expressing an opinion about them. Expert witnesses are generally technical professionals with extensive experience and in-depth knowledge in a particular subject matter. They analyze facts, evidence and information and give testimony on their findings as well as their expert opinions.

Given the diversity of training and experience of chemical engineers, they can be called upon to serve as experts in correspondingly diverse fields, some of which may or may not fall under a classic view of chemical engineering.

Collaboration between insurers, lawyers and technical experts typically involves, but is not limited to, the following:

- Developing a clear understanding of the claims and allegations.
- Knowing and understanding the involved parties and the type of business they are in.
- Gathering the information, evidence and facts necessary for legal, technical and business analyses.
- Performing a screening analysis of the strengths and weaknesses of legal, technical and business factors to support or rebut the claims and allegations.
- Determining the strategies for the approach and steps to be implemented by the legal, technical and business professionals in handling the matter.

3 Selected Examples

Example 1 -- Patent Related Lawsuit

A producer of asphalt shingle roof products hired a retired former senior executive as a consultant to reduce disposal costs of reject materials from production. Roofing shingles typically contain asphalt, filler material and granules evenly placed on a core material such as fiber glass or cellulose. The produced material is cut into standard sizes called "a course." The cutting produces some unusable material ("rejects" or "rejected material").

The consultant researched, developed and pilot-tested a method to recycle rejects back into production. The reject material, containing asphalt and filler material, was finely ground with limestone and then recycled back into production, saving costs of filler material and asphalt and avoiding costs associated with the disposal of the reject material. The tests were successful and easily scaled up to commercial level.

The consultant asked the company to file a patent for the new recycling method, but the company procrastinated and, ultimately, did not file for a patent.

Later, the consultant, after ending his work with the company, filed and obtained a patent for the new recycling method. Upon learning of the patent, the company sued the consultant/inventor for stealing proprietary information, which initially prevented him from offering the technology to others. The viability, cost effectiveness, and worth of the new technology as well as information protection were all issues raised in the lawsuit.

Analysis showed that the proposed retrofit of an average size recycle plant at 70 plants in the industry across the U.S. would roughly total $156,000,000 in capital costs. Estimated annual savings to the industry from reduced raw material costs and avoidance of disposal costs amounted to $190,000,000, yielding an attractive return on investment. So, the viability and cost effectiveness of the new technology were demonstrated in principle.

The worth of the new technology/licensing value was determined by financial and economic experts.

The matter went to trial, and the consultant/inventor won the case.

Principles familiar to technical specialists applied in this example: Manufacturing processes, unit operations, industry survey, capital cost estimation, annual cost estimation, return on investment, and protection of important information.

Example 2 -- Technology Transfer, Design-Build, Project Management Lawsuit

The owner of an operating plant wanted to convert by-product waste to useful and saleable products. The owner acquired a license for new technology to be used for the conversion of waste to intermediates. The owner also acquired established technologies to convert intermediates to saleable products. The new technology licensor represented to the owner that the technology was demonstrated and ready for commercial use. The licensor also offered used equipment in good condition, which the owner agreed to purchase.

A design-build contract for the plant and process for the new technology was awarded on a fixed price ($150 million) and fixed schedule basis to an architect and engineering firm ("A&E").

The technology licensor provided a Process Information Package (PIP) for design-build work as "frozen" --- that is, ready for detailed design and procurement. Numerous and untimely changes were imposed by both the technology licensor and the owner, causing delays in the detailed design, procurement and project completion. The A&E submitted change orders to the owner, who orally approved them.

The technology sold to the owner was actually its first application on a commercial scale in the U.S. The only other reported use of the technology on a commercial scale was in Japan, where several modifications were needed to make the process work correctly. The condition of the used equipment sold to the owner was not "good," as claimed. The equipment required significant cleaning and modification to meet design and code requirements.

Project completion was delayed by three months but, when completed, the system operated satisfactorily. The A&E submitted a Request for Equitable Adjustment ("REA") to the fixed price for the additional costs. In response, the owner sued the A&E for "Bad Project Management" claiming damages amounting to seven times the REA amount. Of note is that the technology licensor was not drawn into the lawsuit and was not held accountable for changes made to the "frozen" PIP.

The matter went to mediation, but did not settle. The owner later filed for bankruptcy, and the plant was bought by another party.

Principles familiar to technical specialists applied in this example: Review of process design, PFDs, P&IDs, impact of design changes and delays on construction, project management, and contractual performance.

Example 3 -- Process Technology Licensing Lawsuit

Two separate companies (A and B) had plants located at one site. Company A manufactured product X1 and sent its wastewater to Company B for treatment and discharge. Company A then started to manufacture another product, X2, at the same plant. Company A recognized that Company B's plant could not handle the additional wastewater created by the process to manufacture X2. Therefore, Company A decided to build its own new treatment plant on the site to handle the wastewater generated by products X1 and X2 combined.

Company A chose a wastewater treatment contractor (TC) to provide "turn-key" services, including technology, engineering, design, construction and startup of the new wastewater treatment plant. The selected technology was a "thermophyllic" biological treatment process, patented and licensed by a third party technology developer ("Licensor"). The license to operate using the technology was granted to Company A, and the technology and engineering details were given by the Licensor to TC.

The Licensor pilot-tested Company A's waste stream to confirm treatability and provided TC with process and engineering information tailored to treat the tested waste stream samples.

At startup, the new wastewater treatment plant failed to meet the treatment targets, and Company A filed a lawsuit against TC. The Licensor was drawn into the lawsuit as a third party defendant.

During the discovery phase of the lawsuit, the allegations against the Licensor were determined to be unfounded both technically and contractually. It was established that there were numerous deficiencies in Company A's design basis as well as defects in both the engineering design and installation by TC. There also were issues involving the performance warranties between Company A and TC.

The technology provided by the Licensor had been commercially proven at other facilities, and it was established that the Licensor had met its contractual obligations.

The matter was settled before trial.

Principles familiar to technical specialists applied in this example: Process evaluation, pilot testing, process design and engineering analysis, contractual commitments, review and understanding of Company A's and TC's obligations and failures.

Example 4 -- Plant Incident and Alleged Medical Injury

A manufacturing plant was undergoing a maintenance turnaround with contractor employees working on scaffolding near the top of a large carbon steel tank containing concentrated sulfuric acid. An undetected water leak into the tank diluted the acid and a relief valve on top of tank momentarily lifted and released vapors with a sulfurous odor. The contractor employees immediately evacuated the area and reported the incident. The turnaround work was shut down for the day, and all workers were sent home.

Reportedly, the contractor employee who was closest to the top of the tank went to a restaurant with his co-workers, where he collapsed. He was attended to and recovered. Later, the contractor employee sued the plant for negligence, alleging that the gas he was exposed to was hydrogen sulfide (H_2S). During the litigation, medical statements were produced stating that the contractor employee's nervous system was permanently damaged from exposure to H_2S.

The defense argued that the released gas was more likely to be sulfur dioxide (SO_2) and not H_2S. Reported public information showed that exposure to SO_2 could have bronchial effects but not permanent nerve damage.

A literature search as well as an examination of the conditions of the carbon steel tank (pressure, temperature, and acid concentration) on the date and time of the incident suggested that the gas released was as a result of corrosion, and it likely contained primarily SO_2 with trace amounts of H_2S.

Defense lawyers engaged an independent third party laboratory to test evolution of gases from reaction of concentrated and diluted sulfuric acid on carbon steel, thus simulating conditions in the tank on the day and time of the release. The tests showed that the evolved gas content was mainly SO_2 with likely trace amounts of H_2S.

A chemistry professor for the plaintiff testified in his deposition that the gas released at the time of the incident was primarily H_2S, but his testimony was strongly challenged by defense lawyers. A second professor from another university testified in his deposition that the plant violated OSHA's Process Safety Management (PSM) regulation. The second professor's testimony was challenged on the basis of his lack of experience with PSM.

Later, it was learned that the contractor employee who filed the lawsuit had been previously employed as a sheet rock worker and had developed indications of dust-related issues involving his bronchial system.

The matter was settled between the parties before trial.

Principles familiar to technical specialists applied in this example: Chemistry of reaction of dilute acid with carbon steel; examination of the incident timeline and possible cause(s) of water backflow into the acid tank; basic metallurgy of carbon steel; review of third party test results; and review of the plant's PSM program.

4 Observations and Conclusion

1. Insurance claims and lawsuits can arise from a variety of reasons and situations, and outcomes are not easily predictable.
2. Lawsuits and insurance claims can arise from unwritten "perceived expectations" by the involved parties.
3. Besides the terms and conditions defined in a contract, other factors, such as standard of care and generally accepted good practice in a given industry, are often encountered in allegations.
4. In addition to directly involved parties, all other entities that are in any way associated with the matter may be drawn into the lawsuit - especially those perceived to have "deep pockets."
5. Claims of advantages and success using new technology on a commercial scale should be thoroughly evaluated before acquiring the license for its use.
6. Clear agreements between a company, its consultants or involved third parties regarding ownership and rights regarding intellectual property need to be executed in advance.
7. Applying fixed price and fixed schedule terms to design-build contracts in capital projects (or any) in which a firm and definitive scope cannot be described clearly in the contractual documents will most likely result in cost overruns and schedule delays, prompting lawsuits.
8. A technical professional can take steps to minimize liability, although it is doubtful one can completely avoid it. Some exemplar suggestions are:
    - Try to clearly understand the technical and (to the extent possible) commercial needs of an assignment.
    - Take on only what you are capable of performing on a sound technical basis. Do not make unsupportable assumptions or try to "wing it."
    - Speak up if you are not the best person to take on the assignment. Have it passed on to someone else who can best perform the work.
    - Keep good written factual records of your work and your communications about the work you are doing.
    - Most of all, be factual and truthful in all you do.

Determining The Liability Of A Fire Starting Product

C. J. Abraham
Scientific Advisory Services Ltd
Great Neck, New York

Abstract

Natural Fire Starter is an ethanol based fuel gel which was designed specifically for wood and wood pellet stoves. It is sold in a plastic bottle with a child proof closure. This product had never been cited in any reported accident until a 7-year old was badly burned when he was allowed, unsupervised, to toast marshmallows in a wood burning fireplace. The subsequent events reported by the child, his parents and brother are discussed as well as the experts and lawyers claiming the liability of the defendants. Said defendants include the manufacturer of the formulation, the bottle manufacturer, manufacturer of the closure and the packager.
In order to reconstruct the accident and determine the cause and liability involved, this investigator had to consider the testimony of the injured child, his parents, brother , the ENT personnel, medical personnel, plus testing and testimony of all of the experts as well as the statements made by the plaintiff's attorney.

Determining The Liability Of A Fire Starting Product

As simple as the product appears, over 50 characteristics and variables of the bottle of fire starter were considered in reconstructing the accident.

After extensive testing by this investigator and all of the experts on both sides, analysis of all of the documents and videotaped depositions, the conclusion of this investigator indicates that the case, as presented by the plaintiff's attorney, is highly suspect and the opinions of the plaintiff's experts are unreliable and without scientific substance.

Background to the accident

On or about February 2010, Plaintiff "A" received burn injuries at his home located in West Virginia. According to the Complaint, he was using a product called Natural Fire Starter Gel, which is an ethanol-based fuel gel.

According to testimony of "A" and his parents, a fire had been burning in the fireplace on the lower level of the house for much of the day. The Natural Fire Starter Gel was near the hearth. At some point in the evening, A was given permission by his mother to go to the lower level of the house unsupervised to roast a marshmallow prior to taking a bath He was naked,.

According to the complaint, A applied the Natural Fire Starter Gel to kindling wood in the fireplace and the Natural Fire Starter Gel container suddenly failed near the lower portion of the bottle, projecting flaming gel onto "A", resulting in injury.

Natural Fire Starter Gel is designed specifically for pellet and wood burning stoves. It is manufactured and assembled by Packaging Service Company (PSC) using a plastic bottle manufactured by C.K.S. Packaging, Inc. and a cap manufactured by Stull Technologies, Inc. The product is distributed by Hearthmark, LLC. The sole purpose of the gelled product was that the product was to be used for only wood pellet and wood burning stoves.

The subject bottle was purchased by Plaintiffs at a local Walmart store. According to the Complaint, Plaintiffs' allege the Natural Fire Starter Gel was not reasonably safe for its intended use, was defectively designed and manufactured, and carried inadequate warnings.

The undersigned reviewed over 10,000 pages of documents obtained through discovery in addition to performing, examining, or analyzing the following:

1. **the subject bottle and exemplars**
2. **Testimony of injured party**
3. **View of fireplace and living room after the accident**
4. **Analysis of the PVC container**
5. **Analysis of the stopper for the PVC container**
6. **Proving or disproving the testimony of the plaintiff, his parents and brother**
7. **Testing and testimony of plaintiff's experts**
8. **Testing and testimony of the defense experts**
9. **Deficiencies in testing by defendant's experts**
10. **Deficiencies in testing by plaintiff's experts**
11. **What are the scientific basis, from a mechanical engineering stand point why either the plaintiff or defense experts were incorrect in their analysis?**
12. **Testimony of all experts.**
13. **Testing by all experts**
14. **Medical documents**

In addition to the above, it was important to evaluate all of the chemical, physical, thermal, and flammable properties of the product in addition to evaluating the warnings and instructions.

1. **Accident Reconstruction**
2. **Annealing**
3. **Basics of blow molding**
4. **Basics o f ethanol based fires**

5. **Basics of gelled fire starters**
6. **Bottle closure**
7. **Brittle versus ductile fracture**
8. **Compatibility of a solvent and chemicals with P VC containers**
9. **Compatibility of ethanol and PVC**
10. **Consumer Product Safety Commission Reports**
11. **CT scans (Computerized Tomography Scan)**
12. **Curvature of bottle at shoulder**
13. **Deflagration, detonation, explosion**
14. **Destructive testing**
15. **Design of PVC container**
16. **Expansion o f PVC plastic**
17. **Explosion limits**
18. **Flame arrestor technology**
19. **Flame speeds**
20. **Flashback**
21. **Flashover**
22. **Heat damage**
23. **Heat exposure and PVC**
24. **Laminar flame speeds of ethanol-air mixtures**
25. **Material Safety Data Sheets (MSDS)**
26. **Napalm**
27. **Non destructive testing**
28. **Origin of fracture**
29. **Pictorials**
30. **Plastic bottle rupturing**
31. **Plasticizers**
32. **Pressure release valves**
33. **Prior reported accidents**
34. **Recalls on subject product and similar products**
35. **Scanning electron microscope**
36. **Seams in plastic container**
37. **Similar products**
38. **Single holes versus multiple holes**
39. **Stains in plastic bottle**
40. **Stress in plastic containers**
41. **Stress cracking**
42. **Stress whitening**
43. **Specification of PVC container**
44. **Supersonic versus subsonic**
45. **T ensile strength of PVC**
46. **The effect of dye in ethanol in PVC bottle**
47. **The effects of gelling ethanol**
48. **Thermal gravimetric analysis**
49. **Viscosity**
50. **Volatility**

51. Warnings and instructions

The Plaintiff, 7 year old boy "A"\ (Deposition summary with comments)

At the time of the incident, "A" was 7 years old and claims in his deposition, that he shoots his own gun. He started shooting at the age of 5. He stated in his deposition that he has never started a fire without an adult present. His mother told him not to use the gel on a burning fire because it was a waste . ***He stated his mother told him that he could start the fire that day. Contrary to that statement*** **"A"'s mother testified that she started the fire prior to taking the family to church.**

"A" stated that he ***"used the gel to start a fire and also to restart a fire."*** **This** ***admission is important and*** **more than documents the fact that "A" squirted a lot more of the Diamond product into the fire than he testified to.** ***On the day of the incident his "mom's back was out and he was sick with pneumonia."***

He stated that he was getting ready to roast marshmallows. However, his mother told him to take a bath. He asked her if he could roast a marshmallow. She said yes. He went to the kitchen, opened up the cabinet ***and got a bag of marshmallows.*** **The alleged bag of marshmallows was not located or identified in any of the documents reviewed or photographs taken at the scene of the accident.**

According to "A", the fireplace was in the basement and his mother, father and brother were upstairs. He walked down the stairs, went to the fireplace and stated that there was no fire . "A" testified that the coals were black with white powder over them, and he thought the fire had gone out. ***If there was white powder all over the coals, how could he see the color of the coals?***

He then stated that he set the marshmallows down and then put one on the stick and laid the stick right by him. He then went and obtained the fire gel. ***He moved the screen and put it on the side.*** **Moving the screen left him without any protection whatsoever. He put some wood up in a teepee and started squirting gel on it. He testified that the gel landed on a coal and started lighting. The heat emanating from the coals is over 400 degrees Fahrenheit and impossible for anyone to expose any part of their body especially an individual's hands. He then stated that he had the bottle up about a foot and the fire went up 6 or 7 inches. This was not accurate based upon the testing performed by the undersigned.**

"A" stated that he held the gel bottle about a foot from the teepee made of wood. He was on his knees andwas holding the bottle and squeezing it angled downward . He then stated thatas he tilted it up and and backward, the gel expelled out of the bottom.He stated it exploded and his brother heard the noise upstairs while he was watching television.

He then stated that he was forced backwards onto the couch and got up and ran into the bathroomon the same floor. He grabbed a cup of water and threw it on theback

of his neck in an attempt to put the fire out. He then ran upstairs to his mother. He stated, contrary to what "T" testified to, that his father grabbed him) and threw him in the bath tub and started running the water.

"A" testified that his mom and dad taught him how to start a fire using a teepee. Contrary to that statement, his father stated under oath that it was "A"'s idea to use the teepee. ***Regardless, starting that type of fire did not include using the gel.***

Although he had observed his mom's friend start a fire using lighter fluid, he has never observed any adult pour lighter fluid on a pre-existing fire.

On the day of the fire, his mother started the fire prior to going to church. The fire was burning most of the day and went out approximately 1 or 2 o'clock in the afternoon. He stated the accident occurred at approximately 8:30 p.m. If that is true and no one replenished the fire with additional logs, it would be questionable if any embers still existed. *Note - How would* **"A" would not be able to build a teepee if the embers were still hot** *and not all ash?* **He stated that there were no coals in the fireplace, there were just ashes. He also stated that matches were "right there." He also stated that if the coals were hot, he would have just put paper on the coals and not used the fire starter gel.** *("A"'s testimony does not make sense.* ***If he didn't use paper and he just observed ashes how could he have built a teepee if there was sufficient heat to ignite the gel? There was a reply by "A" that he did not feel any heat while he made the teepee.***

At the time he was squirting the gel into the fire, he had to stand on the ledge (hearth extension) and lean in to pour the gel onto the wood. He was able to get up on the hearth using his hands and knees. He set the gel down, climbed up on the ledge and started squirting all around the teepee. He ***stated he squirted all around before there was a fire.. He also stated that he was still on his hands and knees when the fire came at him.***

Comment: Reconstruction of the accident by the undersigned with an exemplar fireplace documents that the fire does not flash back *but goes directly in the vertical direction.* **Fire travels directly in the vertical direction.** ***At this time, while on his hands and knees he tilted the bottle back and the bottom came out with a noise.*** **Contrary to the analysis of the plaintiffs' experts, this more than documents that their analysis is without merit and without a scientific basis.**

The amount of squeezes the plaintiffs' expert had to complete prior to meeting within the explosive limits of the ethanol does not coincide with "A"'s testimony or the real-world.

"A" stated that when he first picked up the gel bottle, ***it was not warm. As he fell backwards, he stated that he threw the gel bottle onto the couch.*** *(If there was any leak or opening in the bottom of the subject container, the gel would have* *squirted out the leaking end as well as the opening at the top. "A" never testified that as he was squeezing the gel into the fireplace that any other part of the container was leaking.)*

He had no clothes on when his mother told him to take a bath. However, his older brother, "T", stated that his mother used the wet towel from "A"'s bath that he took prior to roasting the marshmallows in the fireplace.

Further in his testimony he stated that he used a lighter to start fires and would get gel and squirt it on the wood and light it. When he was finished lighting the fire, he would close the cap and set it down about 10 feet from the fireplace. He would then put more wood on the fire. , "A" also stated he was told to be careful of the gel. ***He knew that the gel "could light on fire" and it was dangerous.***

"A" stated that he thought the fireplace was still lit. He had observed that fireplace with the wood burning hours before that time. *(If he thought the fireplace was still burning, why didn't "A" first try paper on the coals as he was taught?)*

He stated that he used the gel even though the ashes were still warm because he wanted to get it started. **He described the ashes as whitish, black and grey. He states that he did not see any red..** *However, he previously testified that the coals were not warm.*

Further on in his testimony on "A" stated that the fire was out about 1 p.m. and he went down to roast marshmallows around 8:30 p.m. (the incident occurred at 7 p.m.). He did not poke a stick in the ashes to see if there were any embers. *(His parents stated that they taught him how to start a fire and use the gel which is not consistent with "A"'s behavior and actions.)* **"A" was allegedly taught how to start a fire and use the gel, which is an unreasonable misuse when it involves a child.**

In addition, based on "A"'s stated awareness and knowledge, and based on the fact his mother taught him not to put the gel on an existing fire or hot embers, "A"'s actions were not reasonable. Moreover, it is not reasonably foreseeable that a 7-year-old child will use the fire starter gel to start or accelerate a fire.

He reiterated the statement he made previously. "A" stated that he squirted the fire starter gel on the hot coal and it flamed up only about 6 or 7 inches. *(However, he previously stated that the coals were not warm.)* **As he tilted the container it came closer to the fire and then "it exploded." This statement cannot be accurate, because the air space contains close to 100% ethanol vapor. The explosive limits within the container are not reached until "A" squirts the gel into the fireplace a number of times so that air can replace the ethanol vapor and bring the concentration of the ethanol to within the explosive limits.** *Hypothetically, assuming, for example, that there was a leak in the bottom seam, deflagration could not*

occur because the opening in the bottom would have released all pressure created by the reaction taking place.

Even though the bottom came closer to the fire, he never saw the flame come in contact with the bottle. "A" reiterated again that once he was on the hearth, he was just on his knees with his hands on the bottle. ***(Note - It would be impossible for the bottom of the container to come near the fire because "A" was squirting into the fire with the nozzle facing the fire and not the bottom of the plastic container. In addition, if "A" was squirting the bottle as he testified to, the gel would have leaked out through the small opening at the bottom if it did exist. There was no testimony by "A" that, as he was squeezing the gel into the fire, that any of the gel came onto any part of his body from the opening in the base of the container.)***

Comment - If "A" was on his knees on the hearth, there should not be any burns below his knees. It would have been impossible for him to be thrown back onto the couch. The burn diagrams from Shriners Hospital show that he is not burned a few inches above both knees and down to his toes in the front of his body except for the inside of his right calf.

The history as recorded by Shriners Hospital for Children states that "patient is a 7 year old white male with recent past history sustained on 08-28-2010. ***The patient had been roasting marshmallows inside when an accelerant for the fire caught on fire and exploded onto the patient. <u>The patient then fell backwards into the fire before being recovered by his father."</u>*** **The areas that were not burned were his scalp, his back and left side. There is no mention of a couch. If he fell backwards into the fire, his back would have been severely burned. It was not. However, it was possible that he fell forward into the fire. That would account for the intense burns on the front part of the body.**

The fact that "A"'s scalp and face were not burned refutes his testimony that he raised the container above his head and it exploded onto his body. If that were accurate, he would have received second and third degree burns on his scalp and his eyes and eye site would have been considerably damaged. ***(This also refutes the plaintiffs' contention that there was a small opening in the bottom of the container prior to the incident.)***

"A"'s mother, in her deposition, stated that she had problems recalling what the warnings stated on the subject container. However, she did remember that the product was flammable based on her experience. If that is true, how could she expect "A" to recall or fully understand any part of what she read and understand the meaning and intent of the warning?

"Father"

"A"'s "Father" also stated that he taught "A" at the age of 5 or 6 to make a fire. He taught him to place a piece of towel or tissue under the small kindling and light it. "A", he stated, came up with the teepee approach. ***(This is contrary to "A"'s testimony.)***

"Father" stated that he used the gel less than 10 times to start a fire and "A" watched him do it. He never saw "A" start a fire with the gel. More importantly, he told "A" that he was never to use the product because it was dangerous. Also, he never put the gel on the mantel and never saw it there. "Father" started the log fires with starter logs.

He purchased only one bottle of the Diamond product at Wal-Mart. That was the only time he ever purchased the product. This comment is contrary to "K"'s comment stating that she purchased it. He stated that he read the label and it said "keep out of the reach of children." That is why he told "A" not to use it. He does recall that it said to "store away from flame or heat sources." That is why he kept it on the floor which, in his opinion, was the coldest point in the room.

It should be noted that, contrary to the plaintiff's allegations, if it was kept on the floor, that would mean that it was well below the hearth and not be exposed to any heat from the fireplace which, the plaintiff's claim caused a leak in the container that ended up causing the fire and the deflagration.

The statement by the father also contradicts "T" who stated that the gel was kept on the mantel. "Father" always let the fire die down in the late afternoon and early evening. He again contradicted "T" when he stated that he tried to smother the fire with a towel.

When he went down to the basement, he does not remember seeing the gel bottle that day or after the fire. He did hear that "A"'s grandmother found the bottle of gel outside. I was under the couch. After the accident, his wife obtained a restraining order preventing him from seeing the children.

"Father"' stated in his deposition where, he quoted an entry in "K"'s (his wife) diary. It is alleged that after she got home on the evening of March 10 and checked the fireplace where the incident occurred, she ***discovered that "A" did not squirt the lighter fluid (*****The gelled product)** ***into the fire as she first thought. The bottle was sitting close to the fire and appeared to have a melted crack near the bottom of the bottle allowing it to spray into the fire causing "A"'s burns. By that time, she had already retained a lawyer to sue.***

Question: "K"'s entry into her diary contradicts everyone else's testimony under oath. Her entry infers that "A" was not on his knees on the hearth holding the container as he testified to under oath. It also contradicts "Father"'s testimony. However, it does fit her attorney's theory and that of

the plaintiff's experts. This author, having over 40 years of experience in evaluating similar cases, the statements made by the family members are incredulous.

"Father" was not aware that "K", his wife at the time, allowed "A" to use the gel under her supervision. He did not expect "A" to start a fire on the evening of the incident. He also stated that before the incident, he thought he heard a popping noise. The question is whether you can actually hear a popping noise in the basement while he is on another floor watching and listening to a TV program.

"K" (mother of "A")
"K" stated n her deposition that "A" should not start fires by himself. She also stated that she never *caught him starting a fire*.

Contrary to "Father"s testimony, she stated that *she bought the fire starter* because it was thick and it was right beside or right near the fire logs in the Wal-Mart store. <u>Without reading the warnings</u>, *she felt that the product was safe and could be used to continue the fire*. She also knew to keep it out of the reach of children and that the product could explode if it got too close to the heat source.(This contradicts "A"' testimony where he stated that his mother read the warnings to him in the store.)

It should be noted that there is not one word about explosions on the Diamond product label. **When "A" and his mother returned from Shriners Hospital in May he told her that *he was squirting the gel on the fire*. She also told aa registered nurse in the emergency department in the hospital that "*The patient squeezed the accelerant into the flames and exploded on the patient*." At her deposition, she stated that she didn't recall telling anyone that statement. However, the statement made and recorded in the hospital notes fit the reconstruction of the cause of "A"'s burns and contradict the statements made by "A" and "K" under oath and also contradicts the opinions set forth by all of the plaintiff's experts.**

She also claimed that she saw a hole in the bottom of the bottle and assumed that it must have sat near the fire because it is only that it could have gotten there, because it got hot and melted.

She does recall seeing the marshmallow on the end of the fork .A small residue of what might have been the remains of a marshmallow was noted in one of the photographs. However, that evidence was never retained for examination. The "bag" of marshmallows were never found.

"K"'s mother found the bottle of the gel outside of the front yard under the couch. There is no explanation how her mother was able to remove the plastic container from under the couch. ***It should be noted that some of the deformation on the subject container was due to the weight of part of the couch on the container caused deformation after the incident.***

She does not recall reading part of the instructions where it stated that one should "Close bottle and remove from immediate area as vapors are flammable." However, she knew that it was flammable.

When asked if she ever read the instructions on the Fire Starter Gel to "A", she replied "I do not recall." However, a question was then asked again of the reading of the label and she stated that she read it to both "A" and "T" and told "A" that he cannot use the product .

"K" stated in an answer to interrogatories that "A" had previously used the Fire Gel) under her supervision). When "K" allowed him to use it under her supervision, there was no fire. His father objected to allowing "A" to use the product. "K" does not recall how many times she allowed "A" to use the product. This contradicts her prior statements under oath. In fact, "A"'s father told "A" not to use the product.

- **"D" (registered nurse)**
  Donielle is a registered nurse working full-time for the emergency department at the Memorial Hospital and the County EMS.

  Her intake states that the patient was roasting marshmallows in the fireplace. *Patient squeezed an accelerant onto flames exploded on patient*. Patient had severe burns over entire body except for his back. Estimated 70% of second and third degree burns observed. No other injuries from explosion.

  This recorded statement contradicts "A"'s testimony, his mother's, his father's and "T"'s.

- **"J" (registered nurse)**
  "J" is an EMT-B. He also testified that he was told that "A" went down to the fireplace and *sprayed some type of aerosol or something on the fire and it blew back on him.*

Shriners Hospital (dictated by "R" MD in April of 1020)
The patient had been roasting marshmallows inside when an accelerant for the fire caught on fire and exploded onto the patient. The patient then fell backwards into the fire before being recovered by his father. **This *more than* documents the fact that "A" sprayed an excessive amount of the Diamond product on to the existing fire. When he fell back, he was exposed to close to 1800 °F. It should also be noted the screen was not in place when this occurred.**

Critique of Dr. "G" (Expert for plaintiff)
Dr. "G"s academic background and experience is excellent. However, his background in storage, transport and processing vessels for chemicals, gasoline and other flammable products in addition to the implementation of various flame suppression and arresting technologies for the safe handling of combustible products

in general and gasoline tank design and leak detection systems for gasoline stations and gasoline storage systems does not relate to the issues and the cause of the accident in the subject case. The "simple screen flame arrester for the inside of a gasoline tank" is not the solution in this specific case. The product in this case is thixotropic and the design of the opening in the cap must be at a size that would allow an individual to squeeze out a desired amount with minimal pressure on the container.

Dr. "G" conveniently forgets about the fact that the plaintiffs testified that "A" was going to roast marshmallows. There was no evidence found of any marshmallows in the area of the fireplace or the living room except for the small amount of white residue on the fireplace fork. There was some white color or remnant on the end of a fork in one of the photographs. However, that was never examined by any of the investigators.

He goes on to state in his report that that "A" "holding the bottle with both hands with the nozzle pointed down and pored the kindling wood when the fire gel hit the kindling/embers and then the heat source ignited the bottle due to the flashback."

Dr. "G" never attempted to reproduce the subject accident with an exemplar fireplace and the subject Diamond formulation. He accepted the questionable testimony of "A" and his mother when designing and performing his testing.

How could **an explosion or flashback cannot occur with one squeeze when the head space contained almost 100% ethanol vapor and testing by the plaintiffs experts documented that a large number of squeezes were necessary to reduce that concentration of ethanol vapor and have sufficient enough air in the head space so that the vapors could be ignited. The experiments performed using an exemplar fireplace by the undersigned disproved all allegations made by the plaintiff and the plaintiff's experts.**

Dr. "G" neglected to state that "A" testified he was on his knees on the hearth and that he allegedly made a teepee out of kindling and squeezed the bottle on the teepee. He also neglected to state what the rest of the family testified as to what occurred as well as other recorded evidence. He also did not mention the fact that there was no evidence of any kindling in the area of the fireplace or the fire. ***If "A" constructed a teepee out of kindling, where was the kindling wood located? Not one of the photographs contained any evidence of kindling wood.***

There is no evidence that the container caught on fire and exploded. If it did, it is more likely than not, that all of the contents would have been burnt up and there would have been extensive deformation of the subject container. His explanation of the deformation of the container is incorrect. The plastic container can be deformed, as it was in this case, when a weight of sufficient size is placed on the plastic for a period of time. The deformation was mechanical and had nothing to do with the incident.

More likely than not, the fire test performed by Dr. "D", another plaintiff's expert , documented that the container had to be placed in a specific position for an extended period of time before it would catch on fire and deflagrate. The amount of time that Dr. "P" had to squeeze the container before anything was notable was well over two dozen times.

Analysis, methodology and opinions (Part of Dr. "G"'s report)

A. "G" Opinion: The rupture to the subject bottle is consistent with an internal explosion.

Comment: The rupture to the subject container is not consistent with an internal explosion. ***The rupture was more consistent with the deflagration. It also required a very large number of squeezes at a specific angle over a long period of time before the fire was spread inside the container before the rupture occurred, and this was not from a "flash back" occurrence. Tests performed by the undersigned with the Diamond container filled with the thixotrophic formulation and refute the experts theory and opinions.***

B. "G" Opinion: The burns on "A" are consistent with burning gel being expelled from the bottom of the ruptured bottle and spraying on his body.

Comment: The burns on "A" are also consistent with "A" spraying an excessive amount of gel in many areas *(into the fireplace and onto his body).* It is also consistent with "A" falling *back* into the fireplace *(see medical documentation).* This can expose "A" up to temperatures reaching up to 1,800 degrees F. No one knows exactly how many times "A" squeezed the plastic container. *For example, it is obvious that he squeezed a large amount of the gel directly in front of the hearth. There is also evidence that there were many squirts all around the fireplace, on areas surrounding the fireplace, the walls and ceilings.* As proven by the undersigned, in all of the experiments performed, no flashback occurs at anytime no matter how much gel is squirted into an open flame.

C. "G" Opinion: Explosion tests were conducted with 5 exemplar bottles of Natural Fire Starter Gel. Ten tests were conducted with single hole caps by squeezing the bottles of gel over a candle flame. In each of these tests, the bottle had an internal explosion. In one of these tests, the bottle ruptured. The ambient temperatures of these tests were between 55 and 65 °F.

Comment: The explosion tests that Dr. "G" conducted on five exemplar bottles did not involve methodology that was scientific. The tests conducted on 8/29 self extinguished. No explosion occurred. The container deflagrated.

The test conducted on 9/11 by "G" where the base of the container blew out did not, in anyway, reproduce the results produced in the exemplar container. The container was held in a horizontal position for an extended period of time which is unrealistic. The same is for the second test that was held on 9/11.. In addition, the initial ignition of the external vapor was snuffed out because the internal concentration of the ethanol was not within the explosive limits.

There is no documentation that the subject container was only partially filled containing a lot of head space. Flammability tests performed by the undersigned using an exemplar fireplace disproved the theory and opinions presented by Dr. "G". The thixotropic Diamond product was squirted directly in front of the fire exposing the spout, vapor and thixotropic liquid to over 200 degrees Fahrenheit and through a hole on the top of the fireplace exposing the spout, vapor and thixotropic liquid to over 300 degrees Fahrenheit. No flashback or internal deflagration occurred while the undersigned personally squeezed the plastic container. The spout was so hot that the undersigned was not able to close it until it cooled down

The additional test *performed by Dr. "G"* with a different cap are also non scientific and do not, in any way simulate the conditions that existed at the time "A" squeezed and squirted the gel all over the pre-existing fire, carpeting and walls.

Using the same formulation with a different cap documented what probably occurred on the original cap on the subject bottle. The subject bottle did not explode but the squirting of the gel all over the fireplace by "A", including his body on a pre-existing fire significantly contributed to "A"'s burns when he was ignited.

<u>If "A" was really going to toast marshmallows, there would have been evidence of marshmallows that were roasted on the pre-existing fire and glowing embers.</u> There was no evidence in the room where the fireplace was located that there was even a supply of marshmallows anywhere else in the house. "A" nor his mother never testified where the marshmallows were originally located. "A" never testified that he took the fireplace fork into any other part of the house and placed a marshmallow or marshmallows on the fork.

D. "G" Opinion: The concentration of the vapor space in the gel bottle was measured with an O2 sensor. This concentration was found to be in the explosive range for temperatures above 50 °F.

Comment: When the cap is first removed the headspace contains well over 19% ethanol vapor. When the container is constantly squeezed, external air constantly dilutes the concentration of ethanol vapor in the headspace to a point where it can fall within the explosive range of 3 to 19%. This can only occur if one squeezes the container a large number of times. That is the only way that the plaintiff's experts were able to create sufficient oxygen in the headspace when the container was constantly squeezed and held in a horizontal angle. When the concentration of ethanol vapor is above 19%, the flame front will be snuffed out.

E. "G" Opinion: The cap on the Natural Fire Starter Gel bottle had on 0.113 inches (2.87 mm) hole and with this cap all of the tests showed internal combustion.

Comment: The statement is not true. Most of the ignition tests were self extinguished *because the concentration of the ethanol vapor was above 19%.* It is only when the container was continually held in the horizontal position on a continuous basis where air was introduced into the air space and the concentration of the ethanol vapor falls within the explosive limits did Dr. "G" obtain internal combustion and not an explosion. If he recalls, "A" testified that he was holding the container in two hands and aiming downward. If that is the case, then there would be no head space and no internal combustion. In the tests that Dr. "G" conducted he should have directed the cap in the downward direction rather than horizontal. In addition, had he placed more gel in the container, it is more likely than not it would have no internal combustion taken place at all.

F. "G" Opinion: Four tests were performed with Kingsford charcoal lighter fluid caps (3 smaller holes) on a Natural Fire Starter Gel bottle. There were no internal combustion for these tests.

Comment: The tests conducted with three small holes were not conclusive and the methodology was not scientific. In addition, the opening in the subject cap was designed to expel a thixotropic formulation. Three holes could have been used for a liquid but not the thixotropic formulation. The use of a three-hole cap would not prevent the underlying incident and would not have acted as a flame arrestor. Moreover, a three-hole cap could expose users to more dangers because if they misuse the product near a flame or hot ember, the consumer has to position the bottle closer than the one with the one-hole cap. In addition, the end of the cap was held much further away from the candle than all of the prior tests with the standard closure.

G. "G" Opinion: The Consumer Product safety committee induced Anderson Chemical to recall 168,000 bottles of Eco-Lite Charcoal starter in 1991. The CPSC learned of five reports of the bottles exploding in consumers' hands when this charcoal lighter fluid was sprayed on hot charcoal. Four of these explosions involved serious injuries, including hospitalizations for burns.

Comment: The recall of the Eco-Lite Charcoal Starter in 1991 had nothing to do with the subject formulation. That product was a liquid formulation not a thixotropic and was designed for a different intended use than the product at issue. A red herring.

H. "G" Opinion: Eco-Lite (82% ethanol, 8% isopranol, 3.5% methanol) is made of a mixture of alcohols and has a flash point of 56 °F. The other more familiar charcoal lighter fluids are made of petroleum products and have a flash point above 100 °F. It should be noted that the Diamond Natural Fire Starter Gel (85-95% ethanol) is made of ethanol and Bitrix and it has a comparable flashpoint to the Eco-Lite mixture.

Comment: What Dr. "G" does not understand is whether the flashpoint was 56 °F or 150 °F, it wouldn't make any difference when the vapor is exposed to an ignition source of 1,500 to 1,800 °F. In addition, the flash point reported was incorrect. The closed cup flash point was significantly higher.

I. "G" Opinion: Caps with 3 small holes are commonly used in the industry and well known. A standard 3 hole cap, manufactured by Stull Technologies, would have prevented this incident as the smaller holes act as a flame arrestor. An igniting flame cannot pass through the small holes and into the container to cause internal combustion. Thus, an explosion cannot occur.

Comment: Dr. "G", in all his tests, did not reproduce the alleged blow (deflagration) out of the subject container in any of his tests. His statement regarding flame arresters is utter nonsense.

The undersigned has been involved with exploding batteries with micro porous flame arresters for over 30 years. The flame front goes through the flame arrester into ullage area much too fast for even the plastic to melt and snuff out the flame front before the battery explodes. The purpose of the flame arrester is to snuff out the flame before it reaches the internal head space of the flammable liquid. When it comes to polymeric materials, the flame front does not melt the opening fast enough to snuff out the flame. It is not foreseeable when one performs a risk analysis on the subject formulation that a foreseeable consumer would constantly squeeze the container when only a half a tablespoon would be used to start a fire. The product was not designed for a 7 year old to use the product as a toy. That is not a foreseeable misuse of the Diamond product.

J. **"G" Opinion: Warnings do not alleviate the product manufacturer and retailer of the duty to design and sell a safe product. It is a basic, generally accepted mechanical engineering principle that performance of a risk assessment (e.g. failure modes and effects analysis) is necessary in product design. It is also a basic, generally accepted mechanical engineering principle to preferentially design out risks as opposed to warn against them. Manufacturers, distributors and resellers have a duty to consider anticipatable uses and misuses of a product.**

Comment: The warnings and instructions on the subject container were well done for a pictorial of a wood pellet stove. ***The product is not designed for 7 year olds. It is designed for a small segment of our population of adults who own wood pellet stoves.*** **It is not designed for fireplaces otherwise a pictorial of a fireplace would have been on the front panel of the product. The wording on the labeling on the container covers that specific area.**

Dr. "G" fails to point out that there were starter logs on the side of the fireplace and that the parents knew that there was a safe method of starting a log fire in their fireplace. He also fails to point out that the parents knew that the subject fire starter was highly flammable and dangerous, must be kept away from children and was designed for wood pellet stoves as shown on the pictorial on the label of the product. The warnings were written for a person with a fifth grade education and would be obvious to any consumer that the small container, like gasoline, should be kept in the garage and out of reach of young children. A manufacturer is not responsible to follow a consumer to their residence and hold their hand while they use their product.

K. **"G" Opinion: It is apparent that none of the defendant's *experts* in this case performed any tests of any kind to determine if the container and cap could be safely used by a consumer in the presence of a flame. Had the management or engineers for these defendants conducted appropriate testing of this product, they would have determined that the Diamond Natural Fire Starter Gel and its container pose a significant and unsafe explosion risk.**

Comment: The defendants did not have to perform any tests because the properties of the gel are known as well as the use and application for pelletized wood stoves. The properties of the one hole cap and its integrity and use for thixotropic formulations is well known for many applications including harmless products such as soap. The same design is also used for squirting insect repelling products inside and outside residential and commercial buildings.

The public also knows that gasoline in any container poses a significant and unsafe explosion risk. The same goes for Drano which is composed of almost 100% of sulfuric acid. *These* products are not designed for 7 year olds.

What Dr. "G" fails to point out is that there have been no reported accidents with fireplaces and the subject product with young children. In addition, there have been no reported accidents with wood pellet heating stoves especially when only a tablespoon of the Diamond product was used to start the fire in a wood or wood pellet stove.

This author measured the force required to release the snap cap of the subject bottle. The tests indicate that it takes over 8-pounds of force to pull the cap off. The cap would be considered child proof. Therefore, it is questionable whether a 7-year opld would be able to easily open the cap with one finger.

L. "G" Opinion: It is also apparent that none of the defendants in this case performed a reasonable risk assessment of this product kind to determine if the container and cap could be safely used by a consumer in the presence of a flame. Had the management or engineers for these defendants conducted an appropriate risk assessment of this product, they would have determined that the Diamond Natural Fire Starter Gel, its cap, and its container pose a significant and unsafe explosion risk. Further they would have discovered the problem had been solved centuries ago and the solution implemented.

Comment: The problem that was allegedly solved centuries ago did not involve 7 year olds. The flammable tests using an exemplar fireplace performed by the undersigned refute all of Dr. "G"'s opinions. In addition, he is not in any position to come to that conclusion without performing his own tests with an exemplar fireplace and the subject Diamond formulation.

M. "G" Opinion: This product as packaged is unreasonably dangerous to be sold as a consumer product.

Comment: The product as sold to the consumer was not unreasonably dangerous because it was not designed or sold with the intent that a 7 year old would be squirting the contents of the bottle on a fire. *In fact, there is no evidence that "A" squirted the contents of the bottle on embers. There is evidence to the contrary that he actually squirted the contents of the bottle onto the pre-existing fire.* In addition, no one knows how many times or how much of the gel that "A" squirted onto the fire once it

got started. He also squirted additional amounts to restart the fire. ***We also know that he squirted the liquid in a large number of other areas. This is illustrative of the fact that "A" was playing with the Diamond product and did not use any of the warnings and instructions his mother stated she gave him previously.***

N. "G" Opinion: An internal explosion can occur in a Diamond Natural Fire Starter Gel bottle as it is being squeezed onto a flame or ember.

The statement that Dr. "G" made has a low probability ***and is totally false.*** **Experiments performed by the undersigned continuously documented that no explosion or deflagration can occur internally no matter how much gel was expelled into the fire from the front or directly on top of the fire. This was documented in the videotape test performed by the undersigned. It is more probable than not that "A" had to squeeze excessive amounts of the gel on the fire in front of him.** ***What no one can answer at this point is when the bottom seam separated and melted. It is more probable than not that the bottom separated well after "A" was on fire.***

In addition, not one of the experts for the plaintiff was able to reproduce the appearance of the subject container in anyway whatsoever. Similarities resulted by squirting the container excessively until sufficient air entered into the head space and reduced the concentration of the ethanol vapor to within the explosive limits. ***This illustration by the plaintiff's experts did not, in any way, simulate "A"'s testimony.***

Squirting the contents excessively into a burning fireplace caused a large flashback onto "A"'s exposed body, ***which had been previously squirted on by "A" himself,*** **to the elevated temperatures emanating from the fireplace while he was** *allegedly* **on his knees facing the unscreened fire.** ***The high temperatures emanating from the fireplace as measured by the undersigned in the tests performed, was more likely than not, much too high for anyone to withstand.***

Contrary to Dr. "G"'s statement, the subject product was not designed to be squeezed on glowing embers or pre-existing fire ***in a fireplace*** **on a continuous basis by an adult or by a 7 year old.** ***The product was created and sold for pellet and wood***

stoves only. This was open and obvious to any adult looking at the label on the container. **If several tablespoons were placed on the glowing embers or pre-existing fire, "A" would not have been injured.**

O. "G" Opinion: This internal explosion can occur when the ambient temperature is above 50 °F.

Comment: Dr. "G"'s tests were not supported by scientific testing and documents readily available in the subject case.

P. "G" Opinion: While the outside temperature was 24 °F, the accident happened in a finished basement of a house where a fire had already been burning to roast marshmallows so the temperature at the time of the incident was likely over 50 °F.

Comment: ***There was no testimony or comment by "A" that, prior to the subject incident, he was able to safely roast and consume a marshmallow.*** **Not one of the experts reported marshmallows being roasted or present in the area. If the experts continually believe the mother and "A", they will continue to make errors and produce results that are not scientifically acceptable. In addition, there was no evidence of a bag of marshmallows available anywhere.**

Q. "G" Opinion: The 5 previous explosions of the Eco-Lite Charcoal starter in 1991 are very similar to the explosion of the present case. The flash point for this previous Eco-Lite product and the present gel are both approximately 56 °F and the chemical compositions of the two products are substantially similar.

Comment: The formulation for the Eco-lite Charcoal Starter in 1991 was a pourable formulation and is not related to the gelled thixotropic formulation. These two formulations should not be intermixed.

R. "G" Opinion: The flashpoint for alcohol is particularly hazardous since at room temperatures, alcohol vapors are in the explosive range. Petroleum distillates like diesel fuel have a higher flashpoint and their vapors tend to be too lean at room temperatures to be explosive. Fuels like gasoline with a very low flashpoint have vapors that tend to be too rich at room temperatures to be explosive.

Comment: When alcohol is placed in a thixotropic formulation, it is much less

dangerous than the pourable liquid. The tests performed by the experts document this statement. Furthermore, the flammable tests performed by the undersigned using an exemplar fireplace and the thixotrophic formulation demonstrated the integrity and the formulation of the container.

S. "G" Opinion: A flame arrestor could prevent internal explosions of this alcohol gel. Flame arrestors have been proven to prevent the propagation of flame into a container. Even a cap with smaller holes, such as the Kingsford cap tested by BEAR, can act as a flame arrestor.

Comment: Dr. "G" failed to test the squirting capabilities of the smaller holes versus the one in the subject container. What he also is demonstrating is that the subject gel does not flow in comparison to a non-gelled liquid. If the product is so highly flammable, it would have emitted sufficient vapors and ignited when first exposed to the radiated temperature emanating from the fireplace.

What Dr. "G" fails to demonstrate is that the open holes have to be small enough to snuff out the flame front entering through the hole openings in the cap, or that the openings will melt and close fast enough to snuff out the flame front before it enters the head space area in the container. That is impossible.

In summary, I am sure that the billing for Dr. "G"'s work conformed to the inadequate testing over a long period of time.

There were many areas that proved that the subject product, when used in accordance with the pictorial that was on the front of the label was relatively safe and not designed to be used by 7 year olds who were suppose to be roasting marshmallows.

More probable than not, based upon over 50 years of experience in the area of flammable materials, fires and the reconstruction of accidents, "A" did a lot more than just squirt the thixotropic product on the fire. There is no doubt that he squirted the product into a number of areas ranging from the carpeting near the hearth, the hearth, the walls and other areas of the living room in addition to squirting a large amount onto a pre-existing fire. The pre-existing fire was a result of "A" starting the fire and then adding the thixotropic gel in large quantities in the area around the fire, the walls, the back of the fireplace and over parts of his body. Dr. "G" ignored the reality of what actually occurred.

Critique "D" (Another PhD plaintiff's expert)

"P" describes the accident stating that ""A" Nelson, a 7 year old boy, was attempting to ignite a fire in a home fireplace when the bottle of fire gel ruptured causing the napalm gel contained in the bottle to spray back onto him resulting in severe burns."

A. “P” states “that the bottle was an unsafe container for a napalm or napalm-like gel.”

Comment - Napalm (naphthenic and palmitic acids) is a thickening/gelling agent generally mixed with gasoline or a similar fuel used in military operations. It is a tactical weapon to remove vegetative cover. *The product at issue is not considered a napalm gel.*

“P”’s conclusion that “the bottle was unsafe” has no scientific or technical basis.”

B. “P” stated “that the bottle was manufactured which is not *fully* compatible with the hazardous contents (i.e., ethanol based napalm gel).”

Comment - PVC containers have been safely used for ethanol and many other solvent and flammable products for generations. The use of the word “fully”, does not, in any way indicate that the contents of the subject bottle are not safely contained for its foreseeable use and application. At room temperature, the container is completely compatible with ethanol if one performs a risk analysis of the foreseeable use and foreseeable misuse of the subject product, the container will not, foreseeably, be exposed to elevated temperatures. The undersigned has personally used PVC plastic containers for ethanol from 1955 to the present time. All of the research performed by the undersigned in graduate school that incorporated pure ethanol was always transferred from large containers from the supplier to small PVC containers. In undergraduate school, the undersigned worked in the chemical supply room for 4 years transferring 80 to 100% ethanol from large containers to a small PVC containers for students to be used for laboratory experiments. The undersigned had the same experience in industry. There was never a question as to the integrity of PVC and its use as a container for ethanol liquid.

Furthermore, this investigator had, in the 90’s patented an ethanol based product in a PVC bottle that contained a non-toxic repellent. Not one accident in the years it was sold throughout the United stated, Europe and China.

C. “P” states “that the bottle wall did not have sufficient wall strength to accommodate the internal forces of an accidental internal vapor ignition.”

Comment - What “P” does not admit was that there is no polymer container that can withstand an internal the deflagration or explosion. Why would he single out this specific polymer and not include every other type of polymer? The flammability experiments performed by the undersigned with an exemplar fireplace and exemplar Diamond containers of the thixotropic formulation refute “P”’s statement.

D. **"P" states "that the bottle wall had too broad a wall thickness variation leading to stress concentration in the thin areas near the bottom."**

Comment - As stated previously, there is no polymer container that can be designed with respect to the subject and similar ones that can withstand internal deflagration and/explosion. However, the flammability experiments performed by the undersigned using an exemplar fireplace and exemplar Diamond containers of the thixotropic formulation refute "P"'s statement.

E. **"P" states "that the thinnest area (weakest and most likely point of rupture) in the wall was near the bottom portion of the bottle which was the area closest to the person holding the bottle."**

Comment - What "P" fails to state or consider is that the head space above the ethanol gel contains close to 100% ethanol vapor or, in the alternative, well over the upper explosive limit of 19%. When the container is turned upside down and a small portion is poured out (a spoonful or two), no change takes place because the thixotropic material moves towards the spout and the head space moves to the rear or bottom of the container. It is only when an individual, such as "A", continually squeezes and squirts the contents many times causing air to then mix with the head space and reduce the concentration of alcohol vapor to within the explosive limits can the vapor ignite and eventually deflagrate internally. ***This was proven and well documented by "P"'s tests. (See reply in I.)***

Furthermore, the differencews in thickness amounted the the thickness of an average human hair.

F. **"P" states "that the appearance of the subject bottle is consistent with the testimony of "A" concerning how the explosion occurred."**

Comment - "P"'s reliance on "A"'s deposition testimony versus everyone else's documented records is incredulous. ***He never attempted to reproduce the subject incident using an exemplar fireplace with the actual product.*** **His testing does not in any way reproduce even "A"'s testimony assuming even part of what he testified to was true. His testimony contradicts the scientific analysis of the ignition and the injuries documented.**

G. **"P" states "there is no way to design and manufacture a squeezable blow-molded plastic container that can safely contain in the detonation of a napalm gel. However, it is possible to design a bottle to handle a deflagration/slow burn without blowing napalm back onto the consumer using the product."**

Comment - The subject container and similar thixotropic products have been safely used by tens of millions of consumers for decades. He continually forgets that the

product was not in any way designed for 7 year olds to play with *or a fireplace*. The product was designed for a small segment of our population that used wood pellet or wood stoves.

Contrary to "P"'s opinion, if there was a practical and inexpensive design for the same container to handle a deflagration, slow burn without blowing "napalm" back onto the consumer it would have been produced and used in lieu of the subject one. If CKS was responsible, and they are not, in manufacturing a plastic container filled with thixotropic ethanol, that would be safe for a 7 year old, it would be supplied in a steel safe with no combination to open it.

If "P" performed a risk analysis relating to the foreseeable use and the foreseeable misuse of the subject product, he would be enlightened to know that the product was designed specifically for pelletized wood burning stoves and not fireplaces. He would also agree that the instructions state that only a tablespoon of the material must be placed within the pelletized wood stove burn pot, that an additional half tablespoon of gel is to be placed on top of the wood pellets, and that the product should not be squirted by a 7 year old all over a room including the fireplace where an excessive amount was squirted continually by "A". In fact, "P" reproduced how many times one has to squeeze the subject container before enough oxygen was drawn into the head space to bring the ethanol vapor concentration down to within the explosive limits. *That is not a foreseeable misuse of the product. It is not foreseeable that parents would allow a 7 year old child to use the product as a toy.*

H. "P" states "Based upon my experience I believe that bottle manufacturers should communicate with their customers to learn the intended use of their bottle. The bottle manufacturer and/or the customer should test the bottle to determine if the bottle is safe for the intended use. If the bottle is found to be unsafe for the intended use, the bottle manufacturer should refuse to market the bottle for that application."

Comment - This more than documents the fact that "P"'s experience is limited in reference to the real world of involving the manufacturing of plastic containers. The sealed area is usually the weakest part of both plastic and glass containers. "P" should also know that if he was capable of performing a risk analysis on the subject container, it would have been easy to *prove that it was not foreseeable that a 7 year old would be playing with the container as a toy*; holding it in a horizontal position adjacent to a flame for an extended period of time and then raising it above his head where he testified it exploded. It should be noted that there is not one other reported incident with the subject product where the incident was described similar to "A"'s.

I. "P" states "a PVC bottle is not appropriate as a container for napalm gel but manufacturing the bottle with the bottom being the weakest point (closest to the user) in the container was exceptionally reckless."

Comment - "P"'s comments again demonstrates his lack of expertise in the area of plastic containers. The seals of all plastic containers are the weakest part of the container. It is also the weakest part of a metal and glass containers. The subject container had seals at the base and on each side of the container. The seal that will be based on how the container was being held while the internal vapors inside the container was on fire. The seals could have failed at any of locations where the seals were located.

The subject plastic container in this case was designed to satisfy load drop-test and also satisfy the internal pressure produced by ethanol vapor. A drop test, in this connection, tests the ability of the bottle to withstand impact forces when the bottle is dropped on its mouth or neck portion because the mouth and/or neck portions are generally dimensionally the smallest parts of the bottle, any impact forces applied to those regions create maximum stresses in the walls of the container that tend to cause the container to burst or rupture at the weakest areas, typically the mold seams. Existing bottles, such as the one in the subject case, used by the chemical industry, have satisfied both requirements or specifications established by the various government authorities and transportation laws. It is important to note that a spherical surface where the seams are located in the subject container is the optimum design for the most resistance to failure.

PVC (Polyvinyl Chloride) is a softer plastic, which is relatively inexpensive. It is very durable and can be dropped without much of a problem. It also takes a lot of deliberate strength to tear PVC apart. The variation of the hardness of PVC is based upon how much of the plastic is put into a specific area. PVC can also be recycled. In addition, PVC has been used to contain ethanol and other alcohol for decades.

"P" also states in his opinion, that PVC is "marginally compatible" with ethanol and that is not true. The rest of his opinion on page 12 of his 28 page report is purely subjective without scientific documentation.

"P" also states that when he was with Dow Chemical, Dow was obligated to work with their customers to insure the safety of the consumer. In this case, CKS was not responsible to perform any risk analysis studies because they supplied the PVC containers as requested. However, he neglects to state that liquid ethanol and other alcohols such as isopropanol have been sold in PVC containers for decades. If that is true, there is no other obligation or responsibility on the part of CKS with reference to the end use application of the PVC container in this case.

If one examines "P"s litigation experience on page 16 and his scientific presentations listed on pages 17-26 as well as the patents listed on pages 28-29, there isn't one reference that relates to PVC, ethanol, fires, explosions and deflagration of containers.

Lastly, the undersigned assumes by choice, that "P" chose not to address the subject of marshmallows even though he had an opportunity to review the depositions of "A" and other members of the family as well as an opportunity to examine all of the photographs associated with the accident. He also failed to comment on the fact that if there was an opening at the base of the container that any squeezing of the container would have released the thixotropic ethanol through the opening in the top closure as well as the alleged opening in the base of the container.

Furthermore, tests performed by the undersigned using an exemplar fireplace and exposing a half filled Diamond product container to well over 200 degrees Fahrenheit documented that the bottom seam would not rupture or melt and that the container actually expanded as shown in the videotapes and photographs taken by the undersigned during and after the tests.

Critique by "H" (Expert for the plaintiff)

Dr. "H" stated in her report that "The product is marketed as a fire-starting accessory, it is suggested for use with wood or pellet stoves."

From Section 3.0 (Background) to page 7 of Dr. "H"'s report, she re-iterates what is already known through prior publications and/or company literature. What is important on page 8 is that she measured head space of the volatiles in the PVC container and found that it was essentially ethanol. There was no report that air and/or oxygen was present in the head space. That is important because no explosion can occur if the concentration of the ethanol vapor is greater than the upper explosion limit.

Dr. "H" refers to the CPSC's 1991 voluntary recall of Eco-Lite charcoal starter. This is a product composed of ethanol and individuals were warned not to spray the material over a burning fire and/or charcoal because, like gasoline it may result in a flashback.

Comment: The objective of the Natural Fire Starter Gel was exactly what the name refers. The name and pictorial on the label designate that the product is designed to be used for wood and pellet stoves and not fireplaces. The warnings Eco-Lite product was not thixotropic and designed to start coal fires for barbeques, as an example. The two products are not designed for the same application.

It is open and obvious to a consumer that one does not squirt gasoline into any type of fire. Therefore, when the consumer is properly warned to keep the product away from children and a parent, as in this case, does not place the subject product out of the reach of their children, including "A", and claims they allow "A" to use the product. Manufacturers cannot be responsible for their actions when they violate the manufactures instructions. Manufacturers cannot be responsible for the parents'

actions when they decide, on their own, to use a product designed for wood and pellet stoves to be used in a fireplace. Dr. "H" ignores that issue.

When a product involves a fire starter of any type, there are no liquid or gelled fire starters that are inherently safe. However, in the subject case, even "A"'s father stated that he used the Safe Fire Starter logs that were on the side of the fireplace. Why any parent would use any other product when they have a safe way of starting a fire is questionable.

What Dr. "H" fails to also point out is that all of the other products do not designate that their products should be used for wood and pellet stoves with a pictorial on the front. *In the opinion of the undersigned, that significantly differentiates the subject product in this case versus all of the others.* The CPSC staff are not warning experts and never addressed that issue with reference to the fact that the Natural Fire Starter Gel with the pictorial and the fact that it was designed only for wood and pellet stoves distinguishes that product from all of the others.

Dr. "H" continually relates the Natural Fire Starter Gel to the Eco-Start Natural Lighter Fuel. They are two separate products designed for two separate objectives. She continually fails to recognize or even discuss the differences in the two and the pictorial of one product versus the other. In fact, on page 12 of her report, she does not feel that there are any differences in the two.

Dr. "H" relates the two by stating that they would act identically with a hot coal or a piece of burning wood "regardless of whether the products are used in wood stoves, barbeque grills, or fireplaces." Dr. "H" is not a warning specialist and, from that stand point alone, *there is a significant difference*.

The Natural Fire Starter Gel is specifically designed for wood stoves and pellets and not barbeque grills or fireplaces. In addition, from a risk analysis standpoint, it is not foreseeable that a parent would use a product designed for wood and pellet stoves or fireplaces when they already had purchased and are experienced in using the Safe Starter Logs. Furthermore, it is not foreseeable that a parent, under supervision or not, would allow a 7 year old to use the product with or without their permission.

"H"'s reports her comments relating to the determination of the subject fuel gel versus what was reported by the manufacturer and the CPSC may vary slightly but it is not an issue in this case. The temperature of the flame of a fire is in the range of 1800 °F and that of burning coals is over 900 °F. What difference would it make whether the flash point of the gel is 54, 73 or 130 degrees? Once the vapor is ignited in the fireplace, it would still ignite and flash back whether the flash point is 52 or 130 degrees.

Therefore, the comments Dr. "H"'s report do not make any sense, and in the opinion of the undersigned, ,the fire starter is safe when it is used in accordance with its instructions. for wood and pellet stoves.

Every consumer has the right to purchase any product they want in Wal-Mart or any other store. For example, many stores sell tanks filled with propane. Propane is highly explosive and has explosive limits between 2.1 and 10.1 percent. Gasoline is also explosive and has explosive limits between 1.4 and 7.6 percent.

When gasoline is used for gasoline lawn mowers, gasoline hedge cutters and emergency electrical power generator, the gasoline is usually dispersed from small container ranging from 1 to 10 gallons. There has never been a recall for any of those products even though they are more dangerous than the thixotropic ethyl alcohol. When the thixotropic ethanol, propane gas and gasoline are used for their intended purpose by adults, the risk of foreseeable injuries is reduced significantly. Dr. "H" fails to continuously point out that all of these products are not designed for parents allowing 7 year olds to use the product under any conditions, *especially when "A" used the product as a toy.*

Lastly, Dr. "H" spent a lot a time reproducing data that is readily available by the manufacturers of the products and documents readily available on line. There was no need to perform any laboratory tests proving what is already known and attempting to reinvent another wheel.

1 "W"J.D.(Expert for the plaintiff)

Mr. "W" defines his area of product safety management as a "system that a reasonable prudent manufacture puts in place before the first product is conceived and ensures that the final product along with its warnings, packaging and marketing materials is reasonably safe."

What Mr. "W" neglected to discuss is that every consumer has the right to purchase any product they want in Wal-Mart or any other store. For example, many stores sell tanks filled with propane. Propane is highly explosive and has explosive limits between 2.1 and 10.1 percent. Gasoline is also explosive and has explosive limits between 1.4 and 7.6 percent.

When gasoline is used for gasoline lawn mowers, gasoline hedge cutters and emergency electrical power generator, the gasoline is usually dispersed from small container ranging from 1 to 10 gallons. There has never been a recall for any of those products even though they are more dangerous than the thixotropic ethyl alcohol. When the thixotropic ethanol, propane gas and gasoline are used for their intended purpose by adults, the risk of foreseeable injuries is reduced significantly. Mr. "W" fails to point out that all of these products are not designed for parents allowing 7 year olds to use the product under any conditions.

The hazards were identified in each of those products; the risks were assessed and adequate safety measures were applied to minimize both hazards. The same approach was made by the manufacturers and distributors of the Natural Fire Starter Gel which is much less dangerous than propone and gasoline. *As stated previously,*

the undersigned has been involved with the use and application of 80 to 100% ethanol and PVC containers for close to 58 years.

It should be noted that Mr. "W" is not a warning specialist especially in the area of hazardous products and chemicals. He points out on page 7 of his report that he is developed on-products warnings and instructions for a number of manufacturers and distributors. Not one of them involved flammable chemicals with physical and chemical characteristics similar to the Natural Fire Starter Gel.

Vending machines, ATVs and pool products are not at the level of warnings and instructions required for products such as the Natural Fire Starter Gel. In contrast, the undersigned has created warnings and instructions for a number of corporations, presented OSHA safety meetings at a large number of construction sites, created Material Safety Data Sheets for flammable products, evaluated polymers from room temperature to hyperthermal temperatures, consulted to NIOSH (the research arm of OSHA) and performed many explosive tests on a variety of flammable products.

Mr. "W" also claims that he created warnings and instructions for fitness equipment in Dick's Sporting Goods. The undersigned has evaluated many accident cases involving fitness equipment. I have also invented sports products that were sold in Dick's. Contrary to Mr. Kitzes' claims, the warnings and instructions for fitness equipment are created by the manufacturers of the equipment and not the management of Dick's Sporting Goods.

Mr. "W"s' report he points out the use of a pictogram but fails to point out the pictorial of a wood stove on Natural Fire Starter Gels label versus that of a fireplace. He also fails to point out that the words in the warning are at the level of a fifth-sixth grader. In addition, he failed to point out that the product must be kept away from children and is not designed to be used by a 7 year old.

The label of the Diamond product states in block letters "PERFECT FOR ALL WOOD AND PELLET STOVES." There is also a pictorial of a wood stove - NO FIREPLACE is shown or mentioned on the front or back labels of the Diamond product. Mr. "W" failed to address that issue which would be considered self-serving on his part by not commenting on that information on the label.

Mr. "W" describes "A"'s behavior prior to the accident without noting all of the different statements made to the EMTs and medical specialists at the hospital. He even failed to mention "A"'s brother's statement under oath where he stated that "A" had already taken a bath and his wet towel was upstairs in the bathroom. In fact, the description of the accident described by Mr. "W" is totally inaccurate and technically incorrect.

Mr. "W" makes a statement regarding "A" being an "avid outdoorsman, hunting, fishing and trapping." What he doesn't point out that "A" at the age of 7 was able to read and did well in school and was sophisticated for his age. *is immature and* *irresponsible, based upon his behavior prior to and up to the time of the incident. In addition, contrary to Mr. "W"'s statement, it should be obvious to him that "K" did not read the Diamond Natural Fire Starter Gel label to "A" on prior occasions. That conclusion is obvious to the undersigned.*

Although Mr. "W" had been associated in some way with the CPSC (not true), he is not a chemist, chemical engineer, safety specialist, safety engineer, professional engineer or certified in warnings and instructions. In addition, he is not qualified to perform a risk analysis, based on his background and experience on the product, Diamond Natural Fire Starter Gel.

The fact that "K", "A"'s mother, ***could not read certain important parts of the label and, on top of that,*** **disregarded the warnings concerning children, negates all of his opinions in regarding the subject warning. When you have a consumer such as "K" dismissing the dangers which she was aware of and the warnings concerning her own child, how could Mr. "W" criticize anything on the label of the subject bottle.**

Mr. "W" knew, or should have known, especially with his law degree, that when you have a recalcitrant worker on a construction job, who refuses to wear protective equipment and is injured, that worker is responsible for his own injuries. In this case, "K" openly refused, in spite of her husband's opinion, to follow the simple instructions and warning on the back label of the Diamond Natural Fire Starter Gel. Now that the horse has run out into the pasture, both Mr. "W" and "K" want to place a lock on the barn door. The undersigned handles similar situations like this each week when the son or daughter of an individual receives a brain injury in soccer and can't go back and play unless they wear his headgear (see www.forcefieldheadbands.com). By that time, it is much too late ***to help the child reduce the risk of a brain injury.***

Although the warnings and instructions on a subject label were accurate based upon the manufacturer's intended use of the product in wood and wood pellet stoves, the instructions regarding children were completely disregarded by "K". If she just obeyed that one warning, "A" would never have been injured.

In spite of "K"'s behavior, Mr. "W" continues to quote various sections of the CPSC without addressing the real issues in the case. For example, on page 12 of his report he describes the vapor "may burn with near invisible flame." What does that mean to a 7 year old? The fact is that the warnings and instructions were never designed for a 7 year old to read and use the product. What Mr. "W" fails to recognize is that "K", the mother of "A", had complete control of the product and completely ignored the warnings.

There was nothing complex about the simple warning concerning children. There is nothing that a manufacturer can do if a consumer, such as "K" refuses to follow and ignores important instructions concerning the safety and welfare of her own child.

On top of all of the instructions that should have been followed if "K" used the product, no one used a tablespoon of the product at any time. If a tablespoon amount was placed on any fire, no injuries would have occurred.

The undersigned has been involved with warnings and instructions regarding all types of chemicals from the early fifties to the present time, from the warnings and instructions published by the Manufacturing Chemical Association to the current ANSI Z 129.1 which is the labeling standard for Hazardous Industrial Chemicals-Precautionary Labeling. It is important to note that this is a voluntary standard and not statutory. All of the other references Mr. "W" makes from do not relate to the real issues in the subject case.

Most important, the manufacturers of the product knew that the product will meet the needs of a small segment of the market which is wood and wood pellet stoves.

"W" states that "the highly flammable ethanol based gel with a nearly invisible flame, could flashback to the bottle, causing it to explode in the hands of the consumer and spread the flaming contents over the body of the user." This statement is totally inaccurate and false. It contradicts the tests performed by "P" and also documents the fact that Mr. "W" is limited with his lack of background in the areas of chemistry, chemical engineering, fires and explosions, deflagrations, flashbacks and why and how the container failed in the instant case. In addition, although Mr. "W" never addressed the subject, it is more likely than not, that no polymeric, glass or metal container could have withstood the deflagration that occurred.

What "W" is attempting to do is validate the defective scientific methodology used by the other plaintiff's experts that did not, in anyway, attempt to reproduce "A"'s description of his accident using an exemplar fireplace and the Diamond thixotropic product. The same group of "experts" chosen by the plaintiff are comparing the same inaccurate models and opinions with one another-but not versus the real world.

Furthermore, Mr. "W" never addressed the subject in which "A" irresponsible actions resulted in him squirting the contents of the Diamond Natural Fire Starter Gel in many areas on the carpet, the hearth, the walls and the firewood while it was burning. In addition he held the container for an extended period of time at a horizontal angle just like "P"'s test demonstrated before the vapor, emanating from the open container, caught on fire resulting in the deflagration of the container. ***This did not happen.***

Mr. "W" deficient background relating to the issues in this case is well beyond his capability of explaining what occurred in this case posing the subject incident and resulting injuries to "A". ***His reliance on the other experts inaccurate and inept attempt to reproduce the questionable testimony of "A" and his family rather than independently prove or disprove his case, defines his lack of***

experience and background in the technical issues associated in the subject case.

Fire and burning tests

Based upon over 50 years of experience in testing the flammability of a variety of solid and liquid chemical formulations in addition to testing the subject formulation in a simulated wood fireplace, it is the opinion of the undersigned, with a reasonable degree of safety engineering and scientific certainty, that "A" did not testify in his deposition as to his actual behavior prior to and up to the time he was injured. There is no question, based upon the testing by the undersigned and the plaintiff's expert's testing, that "A" did more than just squirt a little of the thixotropic alcohol into a burning fire. It is more likely than not that he squirted an excessive amount of the contents not only on the carpeting and the areas surrounding the fireplace such as the hearth, the back wall of the fireplace, the riser to the hearth, the walls, etc., but the straw that broke the camel's back was the excessive amount of formulation squirted on himself and the fire itself for an extended period of time until the vapors emanating from his body ignited due to the fire and heat radiating from the fireplace ending up igniting the vapors.

Experiments by the plaintiff's experts document the fact that an excessively large amount of squirts. The only way the vapors inside the container could be ignited was if "A" constantly squirted the contents out and air continued to enter the head space until there was sufficient air in the container to lower the concentration of the ethanol vapor to within the explosive limit so it could ignite. That was also verified in the video tape submitted by the plaintiff's expert.

In addition, the fact that "A" did not receive burns on the top of his head and back documents the fact that he did not raise the container above his head while the gel was coming out the bottom. That had to occur elsewhere such as on the or around the couch. (The resistance of the container from igniting and burning was well demonstrated in the fire test performed by the undersigned.)

Furthermore, this also verifies the fact that, in the opinion of the undersigned, "A"'s mother, "K", contrary to her sworn testimony under oath, never read the instructions to any of her children and never demonstrated that only a tablespoon amount of the contents should be placed on an unlit piece of wood in a wood stove. If she had, this would also document that "A" did not understand anything she attempted to warn "A" about and, as a 7 year old, he was irresponsible and disobedient.

. ***In addition to the above,*** **the undersigned squirted excessive amounts of the Diamond product into the burning fireplace from the front and from the top exposing the spout to well over 200 degrees Fahrenheit in the front and over 300 degrees Fahrenheit one foot from the flame from the top. All of these tests were videotaped and photographed. In both instances, the undersigned could not touch the closure on**

the top to close it. It was much too hot. In all cases, there was no indication of any flame front entering the container because it would be impossible for that to occur. Therefore, those tests disprove "A"'s account and testimony of what occurred prior to the subject incident.

The undersigned has consulted on several hundred burn cases to both children and adults over the last 44 years. Based on that experience and studies, I am able to convey that experience and observations to the following:

a. Explosions of natural gas, propane, gasoline and other flammable liquids, including ethanol, isopropanol, cause intense heat for a very brief time. Clothing, unless it ignites, is protective in flash burns. Hence, flash burns generally have a distribution involving all exposed skin, with the deepest areas facing the source of ignition. For the most part, flash burns reach progressive layers of the dermis in proportion to the amount and kind fuel that explodes.

b. In contrast to flash injuries, flame burns are invariably deep dermal if not full-thickness because of more prolonged exposure to intense heat.

c. Adding and squirting a voluminous amount of Diamond gel into the fireplace in all directions did not result in any flashback whatsoever. The heat emanating from the fireplace was too high in temperature for any human to withstand. If that is true, where was "A" actually positioned at the time his body caught on fire because of the large amount of gel all over his body? How did he really react once the gel all over his body caught on fire?

Based upon the extensive experiments performed by the undersigned on an exemplar fireplace squirting extensive amounts of gel on a burning fire, there was never a flashback experienced under any conditions. Any testimony regarding flashbacks squirting the Diamond product into a fire cannot be reproduced and would be incorrect.

d. With reference to the subject incident and "A"'s burns, it is highly probable based upon the amount of contents that he squirted into the fire, that he could have easily received a significant amount of his burn injuries from the gel that "A" squirted all over his body prior to getting on his knees in front of the fire.

. **Marshmallows**

There is some evidence of a white residue on a large metal fork in one of the photographs. ***If "A" did roast one or more marshmallows and placed the fork down, this would then lend itself to "A" finishing the marshmallow(s) and then continuing to play with the Diamond product.*** **Squirting of the gel onto the carpet, hearth and other areas around the room, including his own body prior to kneeling before the unscreened fireplace and fire contributed to all of "A"'s burn injuries. All of the vapors emanating from his body was exposed to the heat from the fireplace. The temperature was more than enough to ignite the vapors. In fact, the temperature was more than enough to ignite most other solvents. No one ever found the bag of marshmallows "A" claimed he took downstairs to the fireplace.**

Reconstruction of Subject Accident (Flammability Testing)

The undersigned conducted extensive fire tests in Boynton Beach, Florida by creating a exemplar fireplace using hickory wood chunks, paper and the Diamond Fire Starter Gel to start the fire. The following was observed:

a. **As the fire developed, the undersigned squirted large volumes of the gel onto the fire. No flash fire was observed. What was observed was a fire of higher intensity.**

b. **Large volumes of the Diamond product were squirted into the fire directly from the front. No flash fire was observed. What was observed was a fire of higher intensity.**

c. **A metal surface was covered with the gel and placed near the front of the fireplace. This was done to stimulate the gel covering "A"'s body that he previously squirted on himself. The surface burned with high intensity within a short period of time.**

d. **The surface of the 2" X 6" piece of wood, approximately 16" long was coated with the gel and placed along the floor surface adjacent to the front of the fireplace. Within a short period of time the vapors from the surface ignited and the gel on the surface burned.**

e. **Large volumes of the gel were squirted onto a blazing intense fire in the fireplace from a position directly above the fire and in front of the fire. No flashbacks occurred at any time.** ***The spout was exposed to temperatures ranging from 200 to 300 degrees Fahrenheit. The spout, containing the gel never caught on fire.*** **This contradicts all statements made by the plaintiff's experts and "A" in his sworn testimony.**

f. **The Diamond Fire Starter Gel container was placed directly in front of the fireplace. Nothing initially happened to the container. However, additional tests on the integrity of the container proved that it held up exceptionally well when** exposed to temperatures over 200 degrees Fahrenheit as documented in the videotapes and photographs taken by the undersigned during the flammability and integrity testing of the container. *The container was half filled and the only observation made while the container was exposed to over 200 degrees Fahrenheit was that the body of the container facing the fire expanded as shown in the videotapes and photographs.*

g. The Diamond Fire Starter Gel container with approximately 40 cc's of gel then thrown into the fireplace. There was no explosion and it took an extensive amount of time before the container started to burn documenting the fact that the integrity of the design of the container was more than satisfactory.

h. The undersigned squirted large quantities of the Diamond Natural Fire Starter Gel from the opening in the top of the fireplace shown in the photographs and videotapes. The temperature opening of the top of the fireplace was well over *300* degrees Fahrenheit. It was so hot that the undersigned could not touch the top of the closure and seal the top with my bear hands. The temperature was well over ignition temperature of the ethanol vapor. In spite of that, no internal ignition was observed and no explosion occurred contrary to the testing and comments by Dr. "G" and "P". The latter two plaintiff experts did not, in any way, experimentally demonstrate their incorrect opinions.

Testing the Integrity of the Diamond Natural Fire Starter Gel Plastic Container

The undersigned tested the integrity of a half filled container of the Diamond Natural Fire Starter Gel. The fire was started using the Diamond Natural Fire Starter Gel as the accelerant. Instead of a tablespoon of the material, approximately 1/3 of the container was sprayed on the wood nuggets and the vapors were ignited with a Scripto flame lighter. A continuous videotape was taken and the experiment was also documented with a series of photographs. A half filled container of the Diamond Natural Fire Starter Gel was placed approximately 14 inches away from the fire. The temperature of the heat emanating from the fireplace, at the point where the container was placed, was approximately 200 degrees Fahrenheit. No human being could withstand the temperature.

A half filled container of the Diamond Natural Fire Starter Gel was placed in front of the fireplace for over 30 minutes. After approximately 20 minutes, the area below the headspace that was facing the heat emanating from the fireplace swelled after approximately 25 minutes. There was no failure whatsoever and the bottom seam maintained its integrity contrary to the claims made by the plaintiff's experts. It is the opinion of the undersigned, with a reasonable degree of safety engineering and scientific certainty, that the plastic container is more than suitable for the formulation for the Diamond Natural Fire Starter Gel. In addition, unlike many formulations containing alcohol does not deform while in storage.

An additional important note was that the temperature emanating through the top hole in the cover of the garbage can where the undersigned squirted streams of the gel

into the fire below, was well over *300* degrees Fahrenheit. It was so hot, that the undersigned could not handle the closure for the top of the bottle after squirting the gel through the hole. Contrary to the claims made by the plaintiff's experts, there was no flame front that went through the opening in the spout that entered the internal headspace of the container. The reason was that the concentration of the ethanol vapor was above the upper explosion limit *and that the headspace was at the bottom of the container and not the top. When the container was placed in the upright position, and the spout left open until it cooled down, there was no ignition of the ethanol vapor escaping from the opening in the spout.*

The headspace above the half filled container contained mostly ethanol vapor. When exposed to the excessive heat in front of the exemplar fireplace, the gas expanded and the container responded by expanding as well. Contrary to the *defendants* allegations regarding the seam on the container, it would be predicted to anyone with knowledge of plastic containers, liquids and vapors within the container, that the most vulnerable area of the container when exposed to elevated temperatures would be the headspace within the container and not the bottom seam. *Even though this was foreseeable, it was also proven during the experiment by the undersigned.*

Furthermore, based on the demonstration by the undersigned, the plaintiff's claim that a leaked developed at the bottom seam of the container when the container was exposed in some unexplained way to their fireplace, is an impossibility. If it was exposed to excessive heat from the fireplace, the container would demonstrate an area of expansion above the headspace. The evidence in possession of the plaintiff documented that there was no expansion whatsoever. Therefore, it would be impossible for the accident to occur as explained under oath by "A", members of his family and the reports and experiments released by the plaintiff's experts.

- **The retail package provided sufficient barrier to prevent migration of the product or volatiles from the product.**

- **The Material Safety Data Sheet relating to this product was prepared in a reasonable manner and within all applicable standards.**

- **The delivery of this product, as delivered by PSC to Jarden/Hearthmark, was new and merchantable, free from any defects in workmanship or material, and conformed to applicable specifications, state of technology and the standard of care in the industry. Specifically, it is my opinion to a reasonable degree of engineering and scientific certainty, that the product met the specifications set forth by Jarden Home Brands / Hearthmark in the Performance Sepecifications for the Diamond Fire Starter Gel.**

Tests were conducted by the undersigned to determine the force required to open the closure that snapped in place. All measurements were over 8-pounds of force. That

would make the closure child resistant and very difficult, if not impossible for young children to open it up.

Conclusions

It is the opinion of the undersigned, with a reasonable degree of safety engineering and scientific certainty, that contrary to the allegations set forth by the plaintiffs and the plaintiffs expert's, that the cause of all of "A"'s injuries were directly related to his mother's *willful and intentional* disregard of the safety and welfare of her own son and had nothing to do with the formulation of the Diamond Natural Fire Gel or the integrity of the container. *The fact, at this time she is attempting to expiate her guilt by blaming her negligence on anyone of the defendants is a sham.* The conflicting statements by each member of her family, the EMTs and the medical specialist in the hospital more than document this opinion.

The flammable testing performed by the undersigned refuted and contradicted all of the allegations made by the plaintiff and by all of the plaintiff's experts. All of the testing performed by the plaintiff's experts was not realistic and did not relate, in any way, to the real world and what actually occurred at the time "A" was burned.

Summarily, the undersigned further opines that the Diamond Natural Fire Starter gel product is reasonably safe for its intended uses and reasonably foreseeable misuses, that the actions of PSC, Inc. in communicating with Jarden/Hearthmark and blending/packaging product were within all applicable standards of care, and PSC, Inc. did not cause or contribute to "A"'s incident.

The experiments performed and opinions made by the plaintiff's experts did not represent the actual behavior of "A" prior to and up to the time of the actual incident. All of the experts' opinions were *"validated" by the plaintiff's handpicked group of alleged group of "experts" that compared* based on the results of inaccurate testing and questionable testimony of "A" and his family, *but* not versus the real world and real science. Even so, the Plaintiff's experts still could not replicate the version of events provided by the family.

7. **Opinion**
   a. **With over forty years of experience in the area of flammable fabrics, explosions, fires, children's burn injuries, warnings and instructions I cannot find liability in Diamond's Natural Fire Starter Gel.**
   b. **"A"'s parents have a non-delegable duty to safeguard the safety welfare of "A.". They blatantly ignored that responsibility and allowed him to play unsupervised with a dangerous instrumentality. They are responsible for the consequences of their irresponsible behavior.**
   c. **The undersigned has compassion for "A" and his injuries. However, the liability and the cause of his burn injuries are directly related to the disregard of "A"'s's safety and welfare by his parents.**

C. J. Abraham,, PhD, JD, PE, CPC, FRSC, DEE, FTI, BCFE, DCEE, CChE, CChem has a varied background that can be found at: www.scientificadvisory.com. He also has over 45 patents in a variety of arenas as well. Many of the inventions were commercialized. The latest invention can be found on www.forcefieldheadbands.com

AUTHOR INDEX

American Institute of Chemical Engineers
3 Park Avenue
New York, New York 10016-5991